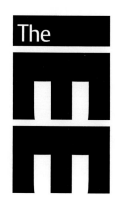

The IEE

D1347956

Guidance Note 7

Special
Locations

17th IEE Wiring Regulations, Seventeenth Edition
BS 7671:2008 C153838926 tallations

Published by The Institution of Engineering and Technology, London, United Kingdom

The Institution of Engineering and Technology is registered as a Charity in England & Wales (no. 211014) and Scotland (no. SCO38698).

The Institution of Engineering and Technology is the new institution formed by the joining together of the IEE (The Institution of Electrical Engineers) and the IIE (The Institution of Incorporated Engineers). The new Institution is the inheritor of the IEE brand and all its products and services, such as this one, which we hope you will find useful. The IEE is a registered trademark of the Institution of Engineering and Technology.

First published 1998 (0 85296 601 6)
Second edition (incorporating Amendment No. 1 to BS 7671:2001) 2003 (0 85296 995 3)
Reprinted (incorporating Amendment No. 2 to BS 7671:2001) 2004
Third edition (incorporating BS 7671:2008) 2009 (978-0-86341-861-7)

Copies of this publication may be obtained from:
PO Box 96
Stevenage
SG1 2SD, UK
Tel: +44 (0)1438 767328
Email: sales@theiet.org
www.theiet.org/publishing/books/wir-reg/

While the author, publisher and contributors believe that the information and guidance given in this work are correct, all parties must rely upon their own skill and judgement when making use of them. The author, publisher and contributors do not assume any liability to anyone for any loss or damage caused by any error or omission in the work, whether such an error or omission is the result of negligence or any other cause. Where reference is made to legislation it is not to be considered as legal advice. Any and all such liability is disclaimed.

ISBN 978-0-86341-861-7

Typeset in the UK by Carnegie Book Production, Lancaster
Printed in the UK by Printwright Ltd, Ipswich

Contents

Chapter 16 Temporary electrical installations for structures, amusement devices and booths at fairgrounds, amusement parks and circuses 123

Cooperating organisations

The Institution of Engineering and Technology acknowledges the contribution made by the following organisations in the preparation of this Guidance Note.

Association of Manufacturers of Domestic Appliances
S.A. MacConnacher BSc CEng MIEE

BEAMA Installation Ltd
Eur Ing M.H. Mullins BA CEng FIEE FIIE
P. Galbraith IEng MIET
P. Sayer IEng MIET GCGI

British Cables Association
C.K. Reed IEng MIET

British Swimming Pools Federation
P. Lang FISPE

City & Guilds of London Institute
H.R. Lovegrove IEng FIET

Department of Energy and Climate Change (DECC)
Eur Ing T. Wickes BSc MSc CEng FEI FEIT

Event Supplier and Services Association (ESSA)
P. Yates MSc MIEE
P. Moule

Electrical Contractors' Association (ECA)
Eur Ing L. Markwell CEng MIEE MCIBSE LCGI
D. Locke BEng (Hons) CEng MIEE MIEEE

GAMBICA Association Ltd
M. Hadley

Health and Safety Executive (HSE)
K. Morton BSc CEng MIEE
N. Gove MEng CEng MIEE

Institution of Engineering and Technology
G.D. Cronshaw IEng FIET (Editor)
P.E. Donnachie BSc CEng FIET
P. Bicheno BSc(Hons) MIET
M. Coles BEng(Hons) MIET
J. Elliott BSc(Hons) PG Cert IEng MIET

Lighting Association
L.C. Barling

National Caravan Council Ltd
J. Lally BEng

Safety Assessment Federation
I. Trueman IEng MIET MSOE MBES

SELECT (Electrical Contractors' Association of Scotland)
D. Millar IEng MIET MILE

Society of Electrical and Mechanical Engineers serving Local Government (SCEME)
C.J. Tanswell CEng MIET MCIBSE

Acknowledgements

References to British Standards, CENELEC Harmonization Documents and International Electrotechnical Commission standards are made with the kind permission of BSI. Complete copies can be obtained by post from:

BSI Customer Services
389 Chiswick High Road
London W4 4AL
Tel: +44 (0)20 8996 9001
Fax: +44 (0)20 8996 7001
Email: orders@bsi-global.com

BSI operates an export advisory service – Technical Help to Exporters (+44 (0)20 8996 7111) – that can advise on the requirements of foreign laws and standards. The BSI also maintains stocks of international and foreign standards, with many English translations. Up-to-date information on BSI standards can be obtained from the BSI website: ww.bsi-global.com

The illustrations within this publication were provided by Rod Farquhar Design: www.rodfarquhar.co.uk

Cover design and illustration were created by The Page Design: www.thepagedesign.co.uk

Copies of Health and Safety Executive documents and approved codes of practice (ACOP) can be obtained from:

HSE Books
PO Box 1999
Sudbury, Suffolk CO10 2WA
Tel: +44 (0)1787 881165
Email: hsebooks@prolog.uk.com
Web: www.hsebooks.com

The HSE website is www.hse.gov.uk

Preface

This Guidance Note is part of a series issued by the Institution of Engineering and Technology to explain and enlarge upon the requirements in BS 7671:2008, the 17th Edition of the IEE Wiring Regulations.

Note that this Guidance Note does not ensure compliance with BS 7671. It is intended to explain some of the requirements of BS 7671, but readers should always consult BS 7671 to satisfy themselves of compliance.

The scope generally follows that of BS 7671; the relevant Regulations and Appendices are noted in the margin. Some Guidance Notes also contain material not included in BS 7671:2008 but which was included in earlier editions of the Wiring Regulations. All of the Guidance Notes contain references to other relevant sources of information.

Electrical installations in the United Kingdom that comply with BS 7671 are likely to satisfy Statutory Regulations such as the Electricity at Work Regulations 1989, however this cannot be guaranteed. It is stressed that it is essential to establish which Statutory and other Regulations apply and to install accordingly. For example, an installation in premises subject to licensing may have requirements different from, or additional to, BS 7671 and these will take precedence.

Introduction

General

This Guidance Note has been revised to align with BS 7671:2008 *IEE Wiring Regulations* 17th Edition. BS 7671:2008 includes additional sections on special locations that were not included in the previous edition, as follows: Marinas (Section 709), Exhibitions, shows and stands (Section 711), Solar photovoltaic (PV) power supply systems (Section 712), Mobile or transportable units (Section 717), Fairgrounds, amusement parks and circuses (Section 740) and Floor and ceiling heating systems (Section 753).

The special locations that were contained in the previous edition of BS 7671 have been revised to align with the latest IEC and CENELEC standards and this Guidance Note has been revised accordingly.

The sections on high protective conductor currents, installation of highway power supplies, street furniture and street located equipment have been removed from this Guidance Note, as these are no longer considered as special locations. The section on extra-low voltage lighting had been removed from this Guidance Note as the requirements are now included in Section 559 of BS 7671:2008.

This Guidance Note provides advice on the special installations and locations of Part 7 (previously Part 6 in BS 7671:2001) of BS 7671:2008, Requirements for Electrical Installations, *IEE Wiring Regulations* 17th Edition, and the special installations and locations for which the International Electrotechnical Commission (IEC) have published requirements.

It is to be noted that BS 7671 and this Guidance Note are concerned with the design, selection, erection, inspection and testing of electrical installations and that these documents may need to be supplemented by the requirements or recommendations of other British Standards.

Other standards of note are described in Regulation 110.1, including BS EN 60079 *Electrical apparatus for explosive gas atmospheres* (other than mines) and BS EN 50281 *Electrical apparatus for use in the presence of combustible dust*.

110.1

The particular requirements of Part 7 of BS 7671 supplement or modify the general requirements contained in the remainder of the standard. Thus, particular protective measures may not be allowed or supplementary measures may be required. However, it is important to remember that in the absence of any comment or requirement in Part 7, the relevant requirements of the rest of the Regulations must be applied.

Section 700

The particular requirements within some Part 7 sections place a prohibition on the use of certain measures of protection, e.g. obstacles and placing out of reach.

International and European Standards

Part 7 of BS 7671 technically aligns with the relevant CENELEC Harmonization Documents (HD) or draft Harmonization Documents (prHD). The preface to BS 7671 identifies the particular CENELEC Harmonization Documents current at the time of publication.

For those persons engaged in work outside the UK the Guidance Note advises whether BS 7671 is based on the European (HD) or the International (IEC) standard.

Contents

This Guidance Note discusses all the sections of Part 7 and includes chapters on special locations/installations not included in BS 7671 as follows:

- ▶ Chapter 9   Medical locations
- ▶ Chapter 13   Gardens
- ▶ Chapter 15   Small-scale embedded generators (SSEG)

The guidance is based on published IEC standards and draft CENELEC proposals, except for Chapter 13 'Gardens'.

Exclusions

The guide does not consider those special installations or equipment where the requirements are specified in other British Standards, such as:

BS EN 60079   ▶ Electrical apparatus for explosive gas atmospheres

BS EN 50281
BS EN 61241   ▶ Electrical apparatus for use in the presence of combustible dust

BS 5266   ▶ Emergency lighting

BS 5839   ▶ Fire detection and alarm systems in buildings.

Locations containing a bath or shower | 1

1.1 Introduction

Section 701 of the 17th Edition includes a number of notable changes.

Section 701

Regulation 701.411.3.3 now requires that additional protection shall be provided for all circuits of the location, by the use of one or more RCDs having the characteristics specified in Regulation 415.1.1. This is a significant change. Previously, only fixed current-using equipment (other than electric showers) located in zone 1 and current-using equipment (other than fixed current-using equipment such as a washing machine, if suitable for use in a bathroom, connected through a fused connection unit) located in zone 3 required 30 mA RCD protection. The updated Regulation 701.411.3.3 means that all circuits, including lighting, electric showers, heated towel rails, etc., will require 30 mA RCD protection. Zone 3 has now been removed.

701.411.3.3

Regulation 415.1.1 specifies the use of RCDs with a rated residual operating current ($I_{\Delta n}$) not exceeding 30 mA and an operating time not exceeding 40 ms at a residual current of 5 $I_{\Delta n}$.

415.1.1

Another significant change is introduced by Regulation 701.512.3. This now permits 230 V socket-outlets to be installed in a room containing a bath or shower provided that they are installed at least 3 m horizontally from the boundary of zone 1. This change removes the ambiguity that existed previously between locations containing a bath or shower and a bedroom containing a shower.

701.512.3

Regulation 701.415.2 introduces a further significant change regarding supplementary equipotential bonding. The regulation states that where the location containing a bath or shower is in a building with a protective equipotential bonding system in accordance with Regulation 411.3.1.2, supplementary equipotential bonding may be omitted, provided that certain requirements are fulfilled.

701.415.2

411.3.1.2

1.2 Scope

The particular requirements of Section 701 apply to locations containing a fixed bath (bath tub) or shower and to the surrounding zones as described in the regulations. The requirements do not apply to emergency facilities in industrial areas or laboratories, on the presumption that they will only be used in an emergency. Where they are used with any regularity the requirements of the section would apply.

701.1

Whilst Section 701 of BS 7671 states that the requirements do not specifically apply to locations containing baths or showers for medical treatment, the Health Authority presently have no special requirements and Section 701 should be applied. However, additional or special requirements may still be necessary for disabled persons.

1.3 The risks

The following information is provided to give a better understanding of why particular requirements are necessary for bathrooms and other wet locations.

Persons in bathrooms are particularly at risk because of a reduction of body impedance due to:

1 lack of clothing, particularly footwear
2 presence of water reducing contact resistance
3 immersion in water, reducing total body resistance
4 ready availability of earthed metal
5 increased contact area.

1.3.1 Clothing
Clothing, particularly footwear such as shoes or boots, can greatly increase the total body resistance.

1.3.2 Body impedance
IEC publication DD IEC/TS 60479-1 *Effects of current on human beings and livestock* provides information on body impedance. Body impedance varies from person to person. The value of impedance depends on a number of factors, in particular on current path, touch voltage, duration of current flow, frequency, degree of moisture of the skin, surface area of contact, pressure exerted and temperature. The IEC document provides information on different total body impedances hand-to-hand for small, medium and large surface areas of contact in dry, water-wet and saltwater-wet conditions. For bathrooms and showers, Table 1.1 shows the varying total body impedance for a current path hand-to-hand, for large surface areas (10 000 mm²) in contact with dry and water-wet conditions for different touch voltages. Large surface areas of contact have the lowest impedances, with medium surface area contact (1 000 mm²) and small surface areas of contact (100 mm²) having impedances of progressively higher magnitudes.

1.3.3 Immersion
Immersion of a body in bath water produces large areas of contact of water and as Table 1.1 shows, this will reduce the total body impedance.

This reduction in body impedance coupled with a location that has earthed metalwork from pipes etc. makes a bathroom particularly hazardous and therefore require special precautions to be taken.

▼ **Table 1.1**   Total body impedances Z_T for a current path hand-to-hand a.c. 50/60 Hz, for large areas of contact in dry and water-wet conditions

| Touch voltage | Values for the total body impedance Z_T (Ω) that are not exceeded for: | | | | | |
| | 5% of population | | 50% of population | | 95% of population | |
(V)	Dry	Water-wet	Dry	Water-wet	Dry	Water-wet
25	1 750	1 175	3 250	2 175	6 100	4 100
50	1 375	1 100	2 500	2 000	4 600	3 675
75	1 125	1 025	2000	1 825	3 600	3 275
100	990	975	1 725	1 675	3 125	2 950
125	900	900	1 550	1 550	2 675	2 675
150	850	850	1 400	1 400	2 350	2 350
175	825	825	1 325	1 325	2 175	2 175
200	800	800	1 275	1 275	2 050	2 050
225	775	775	1 225	1 225	1 900	1 900
400	700	700	950	950	1 275	1 275
500	625	625	850	850	1 150	1 150
700	575	575	775	775	1 050	1 050
1000	575	575	775	775	1 050	1 050
Asymptotic value = internal impedance	575	575	775	775	1 050	1 050

Notes

1   Some measurements indicate that the total body impedance for the current path hand-to-foot is somewhat lower than for a current path hand-to-hand (10% to 30%).

2   For living persons the values of Z_T correspond to a duration of current flow of about 0.1 s. For longer durations Z_T values may decrease (about 10% to 20%) and after complete rupture of the skin Z_T approaches the internal body impedance Z_i.

3   For the standard value of the voltage 230 V (network-system 3N ~ 230/400 V) it may be assumed that the values of the total body impedance are the same as for a touch voltage of 225 V.

4   Values of Z_T are rounded to 25 Ω for dry and water-wet conditions.

1.4   Zones

Zones 0, 1 and 2 provide a very practical method of specifying requirements for protection of equipment against the ingress of water and protection against electric shock, supplementary bonding, etc. in a specific and unambiguous way. Equipment is either in a zone or outside a zone and this can be determined by measurement.

701.32.2
701.32.3
701.32.4

Horizontal or inclined ceilings, walls with or without windows, doors, floors and fixed partitions may be taken into account where these effectively limit the extent of locations containing a bath or shower as well as their zones. There are a number of notable changes in the 17th Edition:

▶ Zone 3 has been omitted. Zone 2 no longer extends above zone 1.

▶ Zone 1 has been extended from 0.6 m in the 16th Edition for showers without a basin for a fixed water outlet to a distance of 1.20 m from the centre point of the water outlet. Demountable shower heads are no longer mentioned.

▶ Zone 1 is now limited by the horizontal plane corresponding to the highest fixed shower head or water outlet or the horizontal plane lying 2.25 m above the finished floor level, whichever is higher.

Examples of the zones for a bath or shower are shown in Figures 1.1a–d.

▼ **Figure 1.1a**
Zones for a bath in elevation view

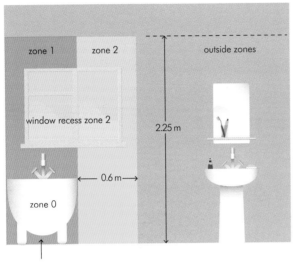

The space under the bath is:
Zone 1 if accessible without the use of a tool
Outside the zones if accessible only with the use of a tool

▼ **Figure 1.1b**
Zones for a bath in plan view

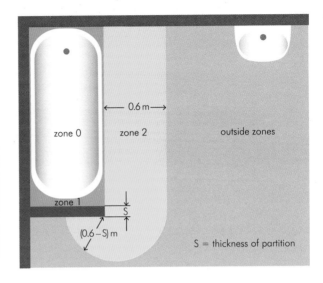

▼ **Figure 1.1c**
Zones for a shower in elevation view

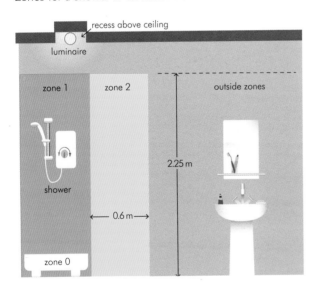

▼ **Figure 1.1d**
Zones for a shower in plan view

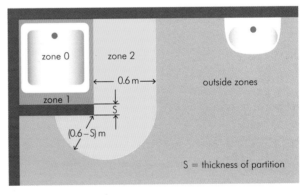

1.5 Protection against electric shock

1.5.1 Locations containing a bath or shower

The 17th Edition includes the following two important changes:

▶ Additional protection shall be provided for all circuits of the location, by the use of one or more RCDs having the characteristics specified in Regulation 415.1.1.
701.411.3.3
415.1.1

▶ Except for SELV socket-outlets complying with Section 414 and shaver supply units complying with BS EN 61558-2-5, socket-outlets are prohibited within a distance of 3 m horizontally from the boundary of zone 1.
701.512.3
BS EN 61558-2-5

The protective measures of obstacles and placing out of reach (Section 417) are not permitted.
701.410.3.5

The protective measures of non-conducting location (Regulation 418.1) and earth-free local equipotential bonding (Regulation 418.2) are not permitted.
701.410.3.6

1.5.2 Supplementary equipotential bonding

Supplementary equipotential bonding is required to be established in all areas within a room containing a bath tub or shower basin, by Regulation 701.415.2 (refer to Figure 1.2). **However, the last paragraph of the regulation allows this bonding to be omitted where the location containing a bath or shower is in a building with a protective equipotential bonding system in accordance with Regulation 411.3.1.2 and provided all of the following conditions are met:**
701.415.2
411.3.1.2

i All final circuits of the location comply with the requirements for automatic disconnection according to Regulation 411.3.2
411.3.2

ii All final circuits of the location have additional protection by means of an RCD in accordance with Regulation 701.411.3.3

iii All extraneous-conductive-parts of the location are effectively connected to the protective equipotential bonding according to Regulation 411.3.1.2.

This means the designer/installer needs to establish that all extraneous-conductive-parts of the location are effectively connected to the protective equipotential bonding according to Regulation 411.3.1.2.

Note: The effectiveness of the connection of extraneous-conductive-parts in the location to the main earthing terminal may be assessed, where necessary, by the application of Regulation 415.2.2.
415.2.2

> **415.2.2** Where doubt exists regarding the effectiveness of supplementary equipotential bonding, it shall be confirmed that the resistance R between simultaneously accessible exposed-conductive-parts and extraneous-conductive-parts fulfils the following condition:
>
> $R \leq 50 \text{ V}/I_a$ in a.c. systems
>
> $R \leq 120 \text{ V}/I_a$ in d.c. systems
>
> where:
>
> I_a is the operating current in amperes of the protective device – for RCDs, $I_{\Delta n}$ for overcurrent devices, the current causing automatic operation in 5 s.

▼ **Figure 1.2**
Supplementary bonding
in a bathroom – metal
pipe installation

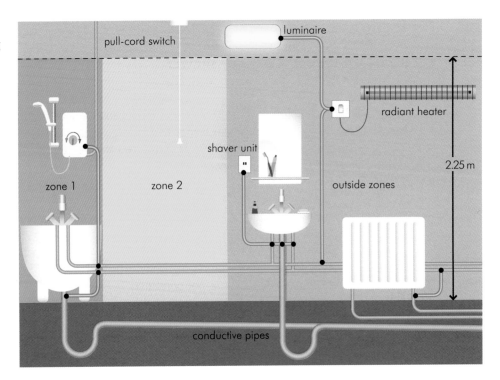

Notes:

1 The protective conductors of all power and lighting points within the zones must be supplementary bonded to all extraneous-conductive-parts in the zones, including metal waste, water and central heating pipes, and metal baths and metal shower basins. Supplementary bonding of circuit protective conductors and extraneous-conductive-parts in the zones does not require multiple connections.

2 Circuit protective conductors may be used as supplementary bonding conductors.

3 The space below the bath/shower is zone 1 if it can be accessed without the use of a tool, otherwise it is outside the zones.

In practice, the resistance between bonded extraneous-conductive-parts and exposed-conductive-parts should not exceed 0.05 ohm.

If supplementary bonding is carried out, the following points are worth noting:

701.415.2   ▶ Supplementary bonding does not necessarily have to be carried out within the bathroom itself but may be carried out in close proximity, such as under the floorboards, above the ceiling, or in an adjacent airing cupboard.

▶ The requirement to supplementary bond to the protective conductors of circuits supplying both Class I and Class II equipment is necessary in case during the life of the installation the user changes a Class II item of equipment for Class I.

1.5.3 Supplementary bonding of plastic pipe installations

Supplementary bonding is not required to metallic parts supplied by plastic pipes such as metal hot and cold water taps supplied from plastic pipes, or to a metal bath not connected to extraneous-conductive-parts such as structural steelwork and where the hot and cold water pipes and the waste are plastic (refer to Figure 1.3).

Supplementary bonding is also not required to short lengths of metal pipe that are often installed for cosmetic reasons when the basic plumbing system is plastic.

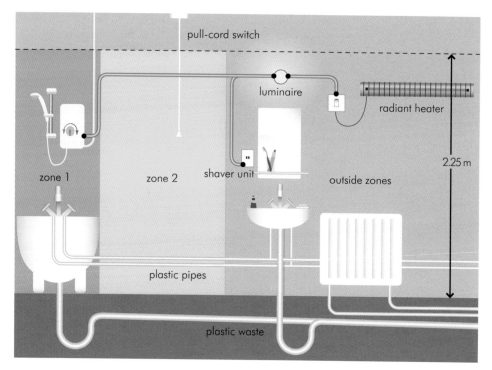

▼ Figure 1.3
Supplementary bonding in a bathroom – plastic pipe installation

Notes:

1   The protective conductors of all power and lighting points within the zones must be supplementary bonded. The bonding connection may be to the earth terminal of a switch or accessory supplying equipment. Supplementary bonding of circuit protective conductors and extraneous-conductive-parts in the zones does not require multiple connections.

2   Circuit protective conductors may be used as supplementary bonding conductors.

3   The space below the bath/shower is zone 1 if it can be accessed without the use of a tool, otherwise it is outside the zones.

1.6   Wiring systems

Metal conduit, metal trunking wiring systems, and metal sheathed cables are allowed in the zones of bathrooms; however, these must be included in any supplementary bonding arrangement. They do not have to be supplying equipment within the zones, but IP ratings within zones 1 and 2 would apply.

701.512.2

1.7   Switchgear and controlgear

Switches and controls other than those which are incorporated in fixed current-using equipment suitable for use in that zone or insulating pull cords of cord-operated switches are not allowed in zones 0, 1 or 2. This means that switches are allowed on showers and fans if the IP rating of the equipment including that of the switch is appropriate for use in the particular zone.

701.512.3

Exceptionally, switches of SELV circuits supplied at a nominal voltage not exceeding 12 V a.c. rms or 30 V ripple-free d.c. with the safety source installed outside zones 0, 1 and 2 may be installed in zone 1, or SELV up to 50 V a.c. rms or 120 V d.c. in zone 2. Also, shaver supply units complying with BS EN 61558-2-5 may be installed in zone 2.

Except for SELV socket-outlets complying with Section 414 and shaver supply units complying with BS EN 61558-2-5, socket-outlets are prohibited within a distance of 3 m horizontally from the boundary of zone 1 (refer to Figure 1.4).

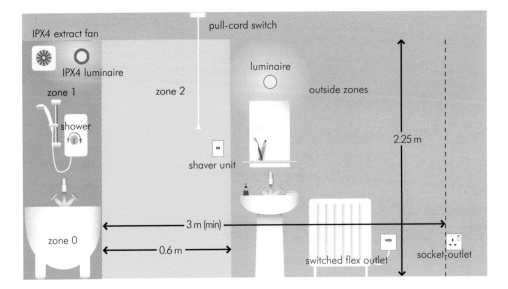

The requirements for switchgear, controlgear and accessories in locations containing a bath or shower are summarised in Table 1.2.

1.7.1 Telephones

Telephones and their sockets should be installed outside zones 0, 1 and 2.

1.8 Current-using equipment

701.55 Fixed and permanently connected current-using equipment may be installed in zones 1, 2 and outside the zones, but there are specific requirements for degrees of protection (see Table 1.2 for summary of the requirements). Equipment having a rated voltage of 230 V may be installed in the above zones provided it has the appropriate IP rating and is suitable for use in the zone. This includes equipment suitable for use in the zone incorporating switches and controls.

Part 2 *Current-using equipment* is equipment that consumes current rather than simply transmits it or switches it. Examples include appliances, luminaires, fans and heaters.

▼ **Table 1.2** Summary of requirements for equipment (current-using and accessories) in locations containing a bath or shower

Zone*	Requirements for equipment in the zones		
	Minimum degree of protection	Current-using equipment, e.g. appliance or luminaire	Switchgear, controlgear and accessories
0	IPX7	Only suitable 12 V a.c. rms or 30 V ripple-free d.c. SELV fixed and permanently connected equipment that cannot be located elsewhere, the safety source being installed outside the zones.	None allowed.
1	IPX4 (Electrical equipment exposed to water jets, e.g. for cleaning purposes, shall have a degree of protection of at least IPX5)	The following fixed and permanently connected current-using equipment if suitable for installation in zone 1 according to the manufacturer's instructions: • Whirlpool units (e.g. hot tubs and whirlpool baths) • Electric showers • Shower pumps • Equipment protected by SELV or PELV at a nominal voltage not exceeding 25 V a.c. rms or 60 V ripple-free d.c., the safety source being installed outside zones 0, 1 and 2 • Ventilation equipment • Towel rails • Water heating appliances • Luminaires	Only switches of SELV circuits supplied at a nominal voltage not exceeding 12 V a.c. rms or 30 V ripple-free d.c. shall be installed, the safety source being outside zones 0, 1 and 2.
2	IPX4 (Electrical equipment exposed to water jets, e.g. for cleaning purposes, shall have a degree of protection of at least IPX5)	Current-using equipment if suitable.	Switchgear, accessories incorporating switches or socket-outlets shall not be installed with the exception of: i switches and socket-outlets of SELV circuits, the safety source being installed outside zones 0, 1 and 2, and ii shaver supply units complying with BS EN 61558-2-5. Note: the IPX4 requirement does not apply to shaver supply units complying with BS EN 61558-2-5 installed in zone 2 and located where direct spray from showers is unlikely.
Outside zones	No additional requirement but subject to Regulation 512.2	No additional restrictions.	Except for SELV socket-outlets complying with Section 414 and shaver supply units complying with BS EN 61558-2-5, socket-outlets are prohibited within a distance of 3 m horizontally from the boundary of zone 1.

* See Figure 1.1 for zones.

1.9 Other equipment, e.g. home laundry equipment

Current-using equipment such as washing machines and tumble-driers is permitted beyond zone 2, subject to the manufacturer's approval. Such equipment must be supplied by means of a permanent connection unit located outside zone 2. Beyond 3 m horizontally from the boundary of zone 1 the equipment may be supplied by means of a plug and socket.

1.10 Electric floor heating systems

701.753 For electric floor heating systems, only heating cables according to relevant product standards or thin sheet flexible heating elements according to the relevant equipment standard may be erected provided that they have either a metal sheath or metal enclosure or a fine mesh metallic grid. The fine mesh metallic grid, metal sheath or metal enclosure must be connected to the protective conductor of the supply circuit. Compliance with the latter requirement is not required if the protective measure SELV is provided for the floor heating system. For electric floor heating systems, the protective measure 'protection by electrical separation' is not permitted.

Section 753 BS 7671:2008 includes particular requirements for floor and ceiling heating systems (Section 753), for which guidance is provided in Chapter 12.

Swimming pools and other basins

2

2.1 Introduction

Section 702 of the 17th Edition includes a number of notable changes, and it now also applies to the basins of fountains.

Section 702

Zones A, B and C in the 16th Edition are replaced by zones 0, 1 and 2 (although the dimensions of the zones remain the same), except that there is no zone 2 for fountains.

A solution is included for the installation of 230 volt luminaires for swimming pools where there is no zone 2 (previously zone C). The 17th Edition states that for swimming pools where there is no zone 2, lighting equipment supplied by other than a SELV source at 12 V a.c rms or 30 V ripple-free d.c. may be installed in zone 1 on a wall or on a ceiling, provided that certain requirements are fulfilled.

Every luminaire must have an enclosure providing Class II or equivalent insulation and providing protection against mechanical impact of medium severity. This is a significant change from the 16th Edition, which only allowed SELV luminaires in zones A and B.

2.2 Scope

702.11

The particular requirements of Section 702 apply to the basins of swimming pools, the basins of fountains and the basins of paddling pools. The particular requirements also apply to the surrounding zones of these basins.

In these areas, in normal use, the risk of electric shock is increased by a reduction in body resistance and contact of the body with Earth potential. Swimming pools within the scope of an equipment standard are outside the scope of the regulations. Special requirements may be necessary for swimming pools for medical purposes. Except for areas especially designed as swimming pools, the requirements do not apply to natural waters, lakes in gravel pits, coastal areas and the like.

For private garden ponds and private garden fountains see Chapter 13.

2.3 The risks

The risk of electric shock is increased in swimming pools and their surrounding zones by the reduction in body resistance (see also bathrooms, section 1.3) and by good contact with Earth arising from wet partially clothed bodies. Equipment installed close to swimming pools and fountains is required to have appropriate degrees of protection against ingress of water.

2.4 Zones

702.32 The requirements for the classification of external influences are based on the dimensions of three zones (examples are given in Figures 702.1 to 702.4 of BS 7671:2008). For a swimming pool, zones 1 and 2 may be limited by fixed partitions having a minimum height of 2.5 m (see Figure 2.1).

▼ **Figure 2.1**
Example of zone dimensions (plan) with fixed partitions of height at least 2.5 m

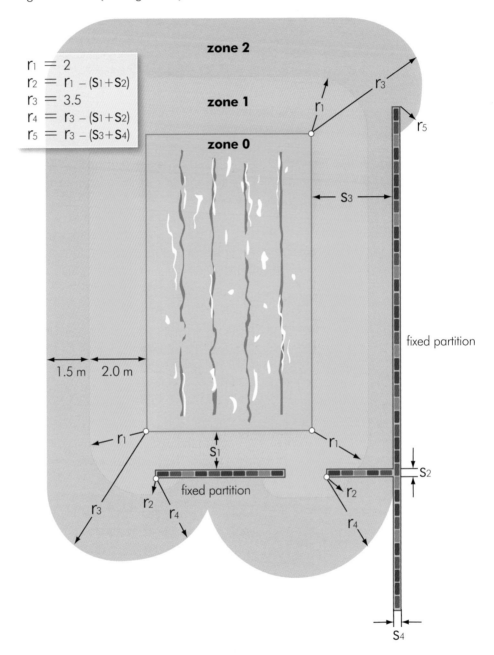

$$r_1 = 2$$
$$r_2 = r_1 - (S_1 + S_2)$$
$$r_3 = 3.5$$
$$r_4 = r_3 - (S_1 + S_2)$$
$$r_5 = r_3 - (S_3 + S_4)$$

2.4.1 Zone 0

This zone is the interior of the basin of the swimming pool or fountain including any recesses in its walls or floors, basins for foot cleaning and water jets or waterfalls and the space below them.

2.4.2 Zone 1

This zone is limited by:

▶ zone 0
▶ a vertical plane 2 m from the rim of the basin
▶ the floor or surface expected to be occupied by persons
▶ the horizontal plane 2.5 m above the floor or the surface expected to be occupied by persons.

Where the swimming pool or fountain contains divingboards, springboards, starting blocks, chutes or other components expected to be occupied by persons, zone 1 comprises the zone limited by:

▶ a vertical plane situated 1.5 m from the periphery of the divingboards, springboards, starting blocks, chutes and other components such as accessible sculptures, viewing bays and decorative basins
▶ the horizontal plane 2.5 m above the highest surface expected to be occupied by persons.

2.4.3 Zone 2

This zone is limited by:

▶ the vertical plane external to zone 1 and a parallel plane 1.5 m from the former
▶ the floor or surface expected to be occupied by persons
▶ the horizontal plane 2.5 m above the floor or surface expected to be occupied by persons.

There is no zone 2 for fountains (see Figure 2.2).

▼ **Figure 2.2** Example of determination of the zones of a fountain

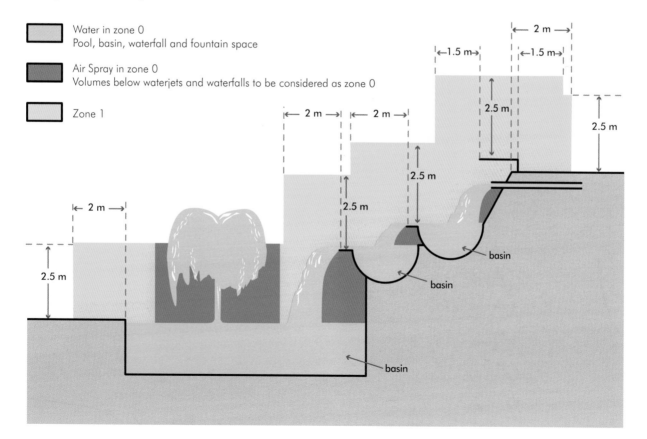

2.5   Protection for safety

702.410.3.5
702.410.3.6   Not surprisingly, the protective measures of obstacles and placing out of reach (Section 417) are not permitted. Also, the protective measures of non-conducting location (Regulation 418.1) and earth-free local equipotential bonding (Regulation 418.2) are not permitted.

702.415.2   Supplementary equipotential bonding is required between all extraneous-conductive-parts and the protective conductors of all exposed-conductive-parts in zones 0, 1 and 2, irrespective of whether they are simultaneously accessible, in accordance with Regulation 415.2. In zones 0, 1 and 2, any metallic sheath or metallic covering of a wiring system (e.g. conduit), either surface-run or embedded in walls, floors etc. at a depth not exceeding 50 mm, must be connected to the supplementary equipotential bonding.

702.52

GN5   See Guidance Note 5 for test methods of determining an extraneous-conductive-part.

702.55.1   There is no particular requirement to install a metal grid in solid floors. However, where there is a metal grid, it must be connected to the local supplementary bonding. It is important to note that Section 753 'Floor and ceiling heating systems' has requirements (such as a metal grid) where electric floor warming is installed.

The requirements for electric floor warming in swimming pool locations would be similar to the requirements for locations containing a bath or shower (see section 1.10).

2.6   Luminaires

2.6.1   Luminaires for swimming pools

702.410.3.4.1   In zones 0 and 1, only protection by SELV is permitted.

In zone 0, only protection by SELV at a nominal voltage not exceeding 12 V a.c. rms or 30 V ripple-free d.c. is permitted, the source for SELV being installed outside zones 0, 1 and 2.

In zone 1, only protection by SELV at a nominal voltage not exceeding 25 V a.c. rms or 60 V ripple-free d.c. is permitted, the source for SELV being installed outside zones 0, 1 and 2.

Note: Equipment for use in the interior of basins which is only intended to be in operation when people are not inside zone 0 shall be supplied by a circuit protected by SELV, automatic disconnection of the supply (Section 411) using a 30 mA RCD, or electrical separation. See Regulation 702.410.3.4.1.

702.410.3.4.3   In zone 2, luminaires are required to be protected either by SELV, a 30 mA RCD or electrical separation. Luminaires can be protected by 30 mA RCDs, but the high protective conductor currents often found in such equipment must be carefully considered. The luminaire manufacturer's advice should be sought with the objective of determining the standing protective conductor current and the maximum protective conductor current during starting so that the luminaires do not cause unwanted tripping of the 30 mA RCD. If RCD protection is used, then the luminaires should be on more than one circuit with separate RCDs.

Where there is no zone 2, section 2.8.1 describes the requirements to be fulfilled to enable 230 V luminaires to be installed in zone 1.

2.6.2 Underwater luminaires for swimming pools
A luminaire for use in the water or in contact with the water must be fixed and must comply with BS EN 60598-2-18.

702.55.2

Underwater lighting located behind watertight portholes, and serviced from behind must comply with the appropriate part of BS EN 60598 and be installed in such a way that no intentional or unintentional conductive connection between any exposed-conductive-part of the underwater luminaires and any conductive parts of the portholes can occur.

2.7   Socket-outlets

Socket-outlets must not be installed in zones 0 or 1, except that for a swimming pool where it is not possible to install a socket-outlet outside zone 1, a socket-outlet that preferably has a non-conducting cover or coverplate, may be installed in zone 1 provided that:

▶   it is installed at least 1.25 m from the border of zone 0 and at least 0.3 m above the floor, and
▶   the supply to the socket-outlet is protected by:
    –   SELV (Section 414), at a nominal voltage not exceeding 25 V a.c. rms or 60 V ripple-free d.c., with the source for SELV being installed outside zones 0 and 1, or
    –   automatic disconnection of supply (Section 411), using a 30 mA RCD having the characteristics specified in Regulation 415.1.1, or
    –   electrical separation (Section 413), supplying only one item of current-using equipment, the source for electrical separation being installed outside zones 0 and 1.

702.53

Socket-outlets may be installed in zone 2 provided the supply circuit is protected by:

i   SELV, with the source being installed outside zones 0, 1 and 2, or
ii   automatic disconnection of supply, using a 30 mA RCD having the characteristics specified in Regulation 415.1.1, or
iii electrical separation, with the source for electrical separation supplying only one socket-outlet and being installed outside zones 0, 1 and 2.

If using SELV or electrical separation to supply a socket-outlet in zone 2, the source can be installed in zone 2 provided the supply circuit is protected by a 30 mA RCD having the characteristics specified in Regulation 415.1.1.

Previously, socket-outlets in these locations had to be an industrial type to BS EN 60309-2 but this is not now a requirement of BS 7671:2008.

2.8   Current-using equipment of swimming pools

702.55.1   In zones 0 and 1, it is only permitted to install fixed current-using equipment specifically designed for use in a swimming pool, in accordance with the requirements of Regulations 702.55.2 and 702.55.4.

Equipment which is intended to be in operation only when people are outside zone 0 may be used in all zones provided that it is supplied by a circuit protected according to Regulation 702.410.3.4.

It is permitted to install an electric heating unit embedded in the floor, provided that it:

i   is protected by SELV (Section 414), the source of SELV being installed outside zones 0, 1 and 2. However, it is permitted to install the source of SELV in zone 2 if its supply circuit is protected by an RCD having the characteristics specified in Regulation 415.1.1, or

ii   incorporates an earthed metallic sheath connected to the supplementary equipotential bonding specified in Regulation 702.415.2 and its supply circuit is additionally protected by an RCD having the characteristics specified in Regulation 415.1.1, or

iii   is covered by an embedded earthed metallic grid connected to the supplementary equipotential bonding specified in Regulation 702.415.2 and its supply circuit is additionally protected by an RCD having the characteristics specified in Regulation 415.1.1.

Note: A socket-outlet and control device of equipment that is intended to be used in the interior of a swimming pool when not occupied should have a notice to warn the user that the equipment should only be used when the swimming pool is not occupied by persons.

2.8.1   Special requirements for the installation of electrical equipment in zone 1 of swimming pools and other basins

702.55.4   Fixed equipment designed for use in swimming pools and other basins (e.g. filtration systems, jet stream pumps) and supplied at low voltage is permitted in zone 1, subject to all the following requirements being met:

▶   The equipment must be located inside an insulating enclosure providing at least Class II or equivalent insulation and providing protection against mechanical impact of medium severity (AG2). This applies irrespective of the classification of the equipment.

▶   The equipment must only be accessible via a hatch (or a door) by means of a key or a tool. The opening of the hatch (or door) must disconnect all live conductors. The supply cable and the main disconnecting means should be installed in a way that provides protection of Class II or equivalent insulation.

▶   The supply circuit of the equipment must be protected by:

–   SELV at a nominal voltage not exceeding 25 V a.c. rms or 60 V ripple-free d.c., the source of SELV being installed outside zones 0, 1 and 2, or

–   an RCD having the characteristics specified in Regulation 415.1.1, or

–   electrical separation (Section 413), the source for electrical separation supplying a single fixed item of current-using equipment and being installed outside zones 0, 1 and 2.

A solution is included for the installation of 230 volt luminaires for swimming pools where there is no zone 2 (previously zone C).

Regulation 702.55.4 states that for swimming pools where there is no zone 2, lighting equipment supplied by other than a SELV source at 12 V a.c. rms or 30 V ripple-free d.c. may be installed in zone 1 on a wall or on a ceiling, provided that the following requirements are fulfilled:

i the circuit is protected by automatic disconnection of the supply and additional protection is provided by an RCD having the characteristics specified in Regulation 415.1.1, and

ii the height from the floor is at least 2 m above the lower limit of zone 1.

In addition, every luminaire shall have an enclosure providing Class II or equivalent insulation and providing protection against mechanical impact of medium severity.

This is a significant change from the 16th Edition, which only allowed SELV luminaires in zones A and B.

2.9　Fire alarms and public address systems

Electrical equipment for safety systems such as fire alarms and public address is likely to be required within the pool area. This equipment should be accessible for maintenance and preferably placed outside zones 0 and 1. Where it is not possible to locate equipment outside zone 1, it should be more than 1.25 metres outside zone 0 and as high above floor level as is practicable, in order to keep the equipment dry. The equipment will need to be of an insulated construction and have IP coding as required by the zone. Local microphones must be SELV or connected via isolating transformers, and telephones, if required, should be cordless-type within the zones with a base unit installed outside the zones.

Reference should be made to the following standards:

▶ BS 5839 – Fire detection and fire alarm systems for buildings
▶ BS EN 54 – Fire detection and fire alarm systems
▶ BS 7445/BS EN 60849 – Sound systems for emergency purposes
▶ BS 7827 – Code of practice for designing, specifying, maintaining and operating emergency sound systems at sports venues.

2.10　Fountains

2.10.1　General

The basins of fountains and their surroundings, unless persons are prevented from gaining access to them without the use of ladders or similar means, are treated as swimming pools.

In zones 0 and 1, one or more of the following protective measures must be used:　　702.410.3.4.2

▶ SELV, the source for SELV being installed outside zones 0 and 1
▶ Automatic disconnection of supply using an RCD having the characteristics specified in Regulation 415.1.1
▶ Electrical separation, the separation source supplying only one item of current-using equipment and installed outside zones 0 and 1.

2.10.2 Electrical equipment of fountains

702.55.3 Electrical equipment in zones 0 or 1 must be provided with mechanical protection to medium severity (AG2), e.g. by use of mesh glass or by grids which can only be removed by the use of a tool.

Luminaires installed in zones 0 or 1 are required to be fixed and must comply with BS EN 60598-2-18.

An electric pump must comply with the requirements of BS EN 60335-2-41.

2.10.3 Additional requirements for the wiring of fountains

702.522.23 Cables supplying equipment in zone 0 should be installed outside the basin, i.e. in zone 1 or beyond where possible, and run to the equipment in the basin by the shortest practicable route.

For cables supplying equipment in zone 0, the designer should check with the supplier that the cable type is suitable for continuous immersion in water.

Cables supplying equipment in zone 1 should be selected, installed and provided with mechanical protection to medium severity (AG2) and the relevant submersion in water depth (AD8). Cable type H07RN8-F (BS 7919) is suitable up to a depth of 10 m of water. For depths greater than 10 m the cable manufacturer should be consulted.

Rooms and cabins containing sauna heaters

3.1 Introduction

While Section 703 'Rooms and cabins containing sauna heaters' retains most of the 16th Edition requirements, there are some notable changes in the 17th Edition.

Section 703

The zones have been simplified and re-designated 1, 2 and 3 in place of the previous A, B, C and D (zones C and D have been combined).

Regarding the supplies to equipment in saunas, additional protection must now be provided for all circuits of the sauna, by means of one or more RCDs having the characteristics specified in Regulation 415.1.1. However, RCD protection need not be provided for the sauna heater itself unless such protection is recommended by the manufacturer, whose advice should be sought.

703.411.3.3

3.2 Scope

The particular requirements of Section 703 apply to:

703.1

i   sauna cabins erected on site, e.g. in a location or in a room
ii the room where the sauna heater is, or the sauna heating appliances are installed. In this case the whole room is considered as the sauna.

The requirements do not apply to prefabricated sauna cabins complying with a relevant equipment standard.

Where facilities such as showers, etc. are installed, the requirements of Section 701 will also apply.

3.3 The risks

There are two particular aspects of saunas that make them special locations:

1   Increased risk of electric shock because of extremely high humidity, lack of clothing, reduced skin resistance and large contact areas
2   Very high temperatures in certain zones.

3.3.1 Zones

The zones are temperature zones, dimensioned down from the ceiling, up from the floor and around the sauna heater. This allows application of the zones whatever the size of the sauna cabin.

703.32
Fig 703

3.4 Shock protection

703.53 Protection against electric shock is provided by not allowing any electrical equipment that is not part of the heating appliance or strictly necessary for the operation of the sauna, such as sauna thermostat, sauna cut-out and luminaires. Light switches must be placed outside the sauna room or cabin and socket-outlets should not be installed in the location containing the sauna heater. It is advisable not to install sockets near the cabin; the same criteria as for swimming pools should be adopted.

703.512.2 Only the sauna heater and equipment belonging to the sauna heater should be installed in zone 1.

703.1 A sauna is often part of a health or fitness suite and may be associated with a swimming pool, showers or bathing facilities. Such premises should be considered as a whole, and it must be borne in mind that the sauna cabin may well be located within the zones of the swimming pool.

The requirements for basic protection and fault protection in saunas are similar to those for bathrooms and swimming pools.

703.410 The protective measures of obstacles and placing out of reach (Section 417) are not permitted. Also, the protective measures of non-conducting location (Regulation 418.1) and earth-free local equipotential bonding (Regulation 418.2) are not permitted.

3.5 Wiring system

703.52 The wiring system should preferably be installed outside the zones, i.e. on the cold side of the thermal insulation. If the wiring system is installed on the warm side of the thermal insulation in zones 1 or 3, it must be heat-resisting. Any metallic sheaths or metallic conduits must not be accessible in normal use.

3.6 Heating elements

703.55 Sauna heating appliances should comply with BS EN 60335-2-53 and be installed in accordance with the manufacturer's instructions.

The heating elements incorporated in a sauna are likely to be metal sheathed. These, unless specified as having waterproof seals, may absorb moisture and cause the operation of a 30 mA RCD, if installed. RCD protection need not be provided for the sauna heater unless such protection is recommended by the manufacturer.

Construction and demolition site installations

4

4.1 Introduction

While Section 704 'Construction and demolition site installations' retains the majority of the 16th Edition requirements, there are some notable changes in the 17th Edition.

Section 704

The 0.2 s maximum disconnection time has been removed and replaced by clear requirements for the supply to a socket-outlet with a rated current up to and including 32 A and a circuit supplying hand-held equipment with a rated current up to and including 32 A.

For automatic disconnection of supply a TN-C-S system is not to be used for the supply to a construction site, except for the supply to a fixed building of the construction site.

704.411.3.1

Section 704 should be read in conjunction with BS 7375 *Code of practice for distribution of electricity on construction and building sites*.

4.2 Scope

Section 704 applies to all sites of construction work including the repair or alteration of existing buildings and demolition work.

704.1

The requirements apply to fixed or movable installations. The regulations do not apply to:

▶ installations covered by the IEC 60621 series 2, where equipment of a similar nature to that used in surface mining applications is involved
▶ installations in administrative locations of construction sites (e.g. offices, cloakrooms, meeting rooms, canteens, restaurants, dormitories, toilets), where the general requirements of Parts 1 to 6 of BS 7671 apply.

4.3 The risks

Construction sites are potentially dangerous in many ways, but only those dangers that are associated with the risks of electric shock or burns are considered here. The risk of electric shock is high on a construction site because:

1 of the possibility of damage to cables and equipment
2 of the wide use of hand tools with trailing leads
3 of the accessibility of many extraneous-conductive-parts, which cannot practically be bonded
4 the works are generally open to the elements.

4.4 Supplies

704.411.3.1 Regulation 704.411.3.1 states that a TN-C-S system shall not be used for the supply to a construction site, except for the supply to a fixed building of the construction site.

The Electricity Safety, Quality and Continuity Regulations 2002 prohibit the use of a TN-C-S system for the supply to a caravan or similar construction.

Where an existing building includes a supply from a TN-C-S system it should be electrically isolated in a secured manner from the construction site supply. The work should be planned in accordance with the Construction (Design and Management) Regulations 2007. Where the existing building supply remains in use in a part of a building that does not have any intended construction work, the existing supply in the area of construction work should be isolated and secured. In addition, the earthing arrangements for the existing supply and construction supply should not be simultaneously accessible.

4.4.1 Supplies to socket-outlets and hand-held equipment

The previous 0.2 s disconnection time has been removed and replaced by the
704.410.3.10 requirements of Regulation 704.410.3.10:

> **704.410.3.10** A circuit supplying a socket-outlet with a rated current up to and including 32 A and any other circuit supplying hand-held electrical equipment with rated current up to and including 32 A shall be protected by:
>
> (i) reduced low voltage (Regulation 411.8), or
> (ii) automatic disconnection of supply (Section 411) with additional protection provided by an RCD having the characteristics specified in Regulation 415.1.1, or
> (iii) electrical separation of circuits (Section 413), each socket-outlet and piece of hand-held electrical equipment being supplied by an individual transformer or by a separate winding of a transformer, or
> (iv) SELV or PELV (Section 414).
>
> Where electrical separation is used, special attention should be paid to the requirements of Regulation 413.3.4.

BS 7671 strongly prefers the reduced low voltage system for portable hand lamps for general use and portable hand tools and local lighting up to 2 kW, while SELV is strongly preferred for portable hand lamps in confined or damp locations.

4.5 Reduced low voltage

110 V reduced low voltage supplies, with the centre point of the secondary winding of the step-down transformer earthed, limit the voltage to Earth to 55 volts for single-phase supplies and 63.5 volts for three-phase.

Limiting the voltage to 55 volts or 63.5 volts between a live conductor and Earth effectively eliminates the risk of dangerous electric shock to exposed-conductive-parts. Figure 4.1 shows single-phase and three-phase reduced low voltage supplies.

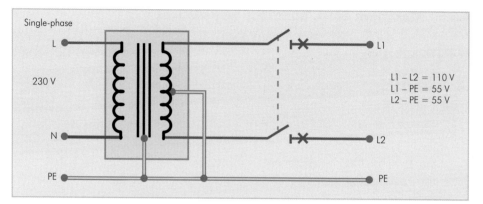

▼ **Figure 4.1**
Reduced low voltage
supplies

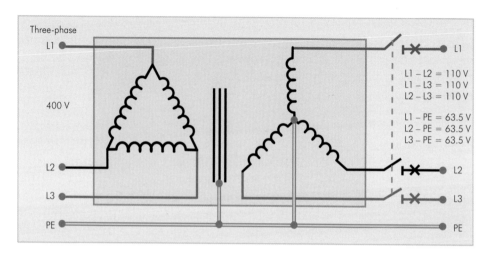

The movable reduced low voltage 110 volt installation equipment is required to comply with BS EN 60439-4. 110 V plugs and sockets to BS EN 60309-2 (or BS 4343) are coloured yellow as recommended by the code of practice for distribution of electricity on construction and building sites (BS 7375). It is usual practice for the cables in the movable installation to have yellow sheaths to identify the voltage being used, although this is not a requirement of BS 7671 or BS 7375. Cable manufacturers produce arctic grade yellow PVC sheathed flexible cables for use in low temperature situations to distinguish from normal PVC sheathed cables.

704.511.1

It should be noted that whilst BS 7671 strongly prefers reduced low voltage systems for portable hand lamps for general use and portable hand tools and local lighting up to 2 kW, the Health and Safety Executive have advised that the use of 230 volt hand tools protected by an individual 30 mA RCD may be acceptable (see HSE publication HSG141 *Electrical safety on construction sites*).

The notes of guidance to the Electricity Safety, Quality and Continuity Regulations advise that special consideration be given to the earthing and protection arrangements. Guidance Note 5: *Protection Against Electric Shock* includes guidance on earthing.

GN5

In addition to reduced low voltage, BS 7375 also provides information on typical distribution voltages for particular applications. For example, 400 V three-phase for fixed or movable plant over 3.75 kW and 230 V single-phase for fixed floodlighting and site building installations.

4.5.1 Maximum earth fault loop impedances

411.8.3 The value of earth fault loop impedance at every point of utilisation, including socket-outlets, must be such that the disconnection time does not exceed 5 s.

The maximum value of earth fault loop impedance is determined by the formula:

$$Z_S \times I_a \leq U_0$$

where:

Z_S is the earth fault loop impedance

I_a is the current in amperes causing automatic operation of the device within the specified time

U_0 is the nominal line voltage to Earth in volts.

Alternatively, if the nominal voltage U_0 is 55 V single-phase or 63.5 V three-phase then Table 41.6 of BS 7671:2008 can be used to determine the maximum earth fault loop impedances for various circuit-breaker types and ratings. However, it should be noted that the values given are not to be exceeded when the conductors are at their normal operating temperature.

If general purpose fuses to BS 88-2.2 or BS 88-6 are used and the nominal voltage U_0 is 55 V single-phase or 63.5 V three-phase then Table 41.6 of BS 7671:2008 can again be used to determine the applicable maximum earth fault loop impedances.

For a fuse of a different type and rating, the appropriate British Standard should be referenced to determine the value of I_a for the disconnection time in accordance with the appropriate value of the nominal voltage U_0. This can then be used to determine the maximum earth loop impedance to comply with $Z_S \times I_a \leq U_0$.

4.6 Wiring systems

704.52 Cables on a construction site location should preferably not be installed across walkways or site roads as they are susceptible to mechanical damage. If cables are installed in this manner they would require the appropriate level of mechanical protection.

For reduced low voltage systems flexible thermoplastic cables rated at 300/500 V and suitable for low temperature (BS 7919) should be used. These cables remain flexible at lower temperatures than standard PVC cables and are ideal for outdoor use. They are referred to as arctic grade cable and typically have yellow (refer to section 4.5) or blue sheaths.

For cables used for applications exceeding reduced low voltage, flexible cables rated at 450/750 V that are resistant to abrasion and water should be used, type H07RN-F (BS 7019) or equivalent. These are heavy duty rubber insulated and sheathed flexible cables suitable for outdoor use.

4.7 Isolation and switching

704.537.2.2 Regulation 704.537.2.2 requires each assembly for construction sites (ACS) to incorporate suitable devices for the switching and isolation of the incoming supply.

A device for isolating the incoming supply must be suitable for securing in the off position, e.g. by padlock or location of the device inside a lockable enclosure.

Current-using equipment shall be supplied by ACSs (see Figure 4.2), each ACS comprising:

i overcurrent protective devices, and
ii devices affording fault protection, and
iii socket-outlets, if required.

Safety and standby supplies must be connected by means of devices arranged to prevent interconnection of the different supplies.

▼ **Figure 4.2**
Examples of assemblies for construction sites

a Typical power assembly

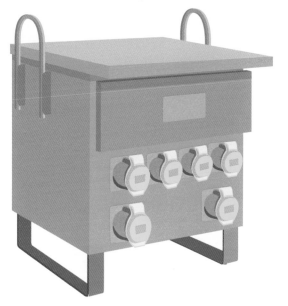

b Typical lighting assembly

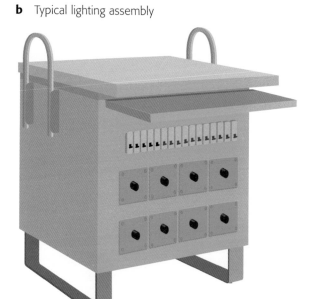

4.8 Protection against the weather and dust

All equipment that is part of the movable installation should have a degree of protection appropriate to the external influences. Equipment for external use should be at least IP44. However, equipment installed in a weather protected location, such as an office being refurbished, would have no specific IP requirement.

4.9 Inspection and testing

It is recommended that the maximum period between inspections of construction site installations is 3 months.

GN3

Fixed installation RCDs should additionally be tested daily (using the integral test button). Should RCDs be used to protect mobile equipment they must be tested by the operative before each period of use (again using the integral test button) and by the responsible person every 3 months (using an RCD tester).

Recommended intervals for inspection and testing are given in Table 4.1.

▼ Table 4.1
Frequency of
inspection and testing
of equipment on
construction sites

Type of equipment (note 1)	User checks (note 2)	Class I		Class II (note 4)	
		Formal visual inspection (note 3)	Combined inspection and testing (note 5)	Formal visual inspection (note 3)	Combined inspection and testing (note 5)
S	None	1 month	3 months	1 month	3 months
IT	None	1 month	3 months	1 month	3 months
M (note 6)	weekly	1 month	3 months	1 month	3 months
P (note 6)	weekly	1 month	3 months	1 month	3 months
H (note 6)	weekly	1 month	3 months	1 month	3 months

Notes:

1   S   Stationary equipment

   IT   Information technology equipment

   M   Movable equipment

   P   Portable equipment

   H   Hand-held equipment

2   User checks are not recorded unless a fault is found.

3   The formal visual inspection may form part of the combined inspection and tests when they coincide, and must be recorded.

4   If class of equipment is not known, it must be tested as Class I.

5   The results of combined inspections and test are recorded.

6   30 mA RCDs supplying this equipment should be tested daily using the integral test button. Before carrying out the test it should be confirmed that it is safe to do so.

Agricultural and horticultural premises

5

5.1 Introduction

Section 705 'Agricultural and horticultural premises' in the 17th Edition has been restructured and there are a number of editorial and technical changes.

Section 705

For automatic disconnection of supply the 0.2 s maximum disconnection time has been removed and replaced by the requirement for the use of RCDs as disconnection devices for circuits, irrespective of the type of earthing system.

Regulation 705.415.2.1 expands the requirements of Regulation 415.2 concerning supplementary equipotential bonding and refers to Figure 705 of BS 7671:2008 at the end of the section showing the requirements for bonding in cattle sheds. A note to this regulation indicates that TN-C-S is not recommended if there is no metal grid laid in the floor.

Under 'Safety services', a new requirement in Regulation 705.560.6 requires consideration of automatic life support in high density livestock environments, to ensure that supplies of water, food, ventilation and lighting are maintained in the event of power failure by means of, for example, an alternative source or back-up supply.

5.2 Scope

The particular requirements of Section 705 apply to fixed electrical installations indoors and outdoors in agricultural and horticultural premises. Some of the requirements are also applicable to other locations that are in common buildings belonging to the agricultural and horticultural premises. Where special requirements also apply to residences and other locations in such common buildings this is stated in the text of the relevant regulations.

Rooms, locations and areas for household applications and similar are not covered by this section.

Note: Chapter 5 does not cover electric fence installations, for which reference should be made to BS EN 60335-2-76.

5.2.1 Residences and other locations belonging to agricultural and horticultural premises

Part 2   These are defined as:

Residences and other locations which have a conductive connection to the agricultural and horticultural premises by either protective conductors of the same installation or by extraneous-conductive-parts.

Note: Examples of other locations include offices, social rooms, machine-halls, workrooms, garages and shops.

5.3   The risks

The particular risks associated with farms and horticultural premises are:

1   general accessibility of extraneous-conductive-parts
2   an onerous environment with respect to mechanical damage, exposure to the weather, corrosive effects (from water, animal urine, farm chemicals etc.)
3   a mechanically hazardous area due to electromechanical equipment, mills and mixers, and mechanical drives of all kinds
4   rodent damage to (gnawing of) cables, leading to fire risks
5   storage of flammable materials, e.g. straw and grain
6   increased susceptibility of electric shock for livestock.

5.4   Electricity supplies

Because of the practical difficulties in bonding all accessible extraneous-conductive-parts electricity distribution companies might not provide a PME earth to agricultural or horticultural installations.

The Department of Energy and Climate Change (DECC), formerly the DTI, guidance on the Electricity Safety, Quality and Continuity Regulations 2002 advises {9(4)} that special consideration should be given to the earthing and bonding requirements to farms where it may prove difficult to attach and maintain all the necessary equipotential bonding connections for a PME supply.

705.415.2.1   Regulation 705.415.2.1 refers to Figure 705 which shows the requirements for bonding in cattle sheds. TN-C-S is **not** recommended if there is no metal grid laid in the floor.

A TN-S supply is unlikely to be provided by a distributor as a routine unless the installation is particularly large and early application is made. Alternatively, consideration should be given to installing an additional earth electrode, as it is most likely that the installation will be required to be TT.

It should be noted that in TT installations isolators are required to switch all live conductors including the neutral.

5.5 Protection against electric shock

The protective measures of obstacles and placing out of reach (Section 417) are not permitted.

705.410.3.5

The protective measures of non-conducting location (Regulation 418.1) and earth-free local equipotential bonding (Regulation 418.2) are not permitted.

705.410.3.6

5.5.1 Protective measure: automatic disconnection of supply

The 0.2 s maximum disconnection time has been removed and replaced by the use of RCDs for automatic disconnection of supply as follows (see Figure 5.1):

705.411.1

> **705.411.1** In circuits, whatever the type of earthing system, the following disconnection devices shall be provided:
>
> (i) In final circuits supplying socket-outlets with rated current not exceeding 32 A, an RCD with a rated residual operating current not exceeding 30 mA
> (ii) In final circuits supplying socket-outlets with rated current more than 32 A, an RCD with a rated residual operating current not exceeding 100 mA
> (iii) In all other circuits, RCDs with a rated residual operating current not exceeding 300 mA.

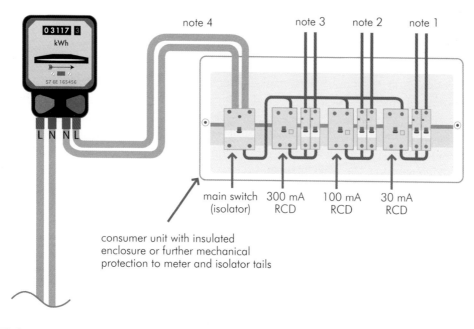

▼ **Figure 5.1**
Requirements for RCDs

note 4   note 3   note 2   note 1

main switch   300 mA   100 mA   30 mA
(isolator)   RCD   RCD   RCD

consumer unit with insulated
enclosure or further mechanical
protection to meter and isolator tails

Notes:

1 Socket-outlet circuits not exceeding 32 A
2 Socket-outlet circuits exceeding 32 A
3 Circuits other than socket-outlets
4 The risk of faults to exposed-conductive-parts on the supply side of the main switch must be minimised as such faults are not detected by the RCDs.

PD 6519-3:1999
(IEC 60479-3:1998)

A BSI published document PD 6519-3 (IEC 60479-3) *Guide to effects of current on human beings and livestock – Part 3 Effects of current passing through the body of livestock* is available that provides guidance on the effects of electric currents on livestock. The report indicates values for the impedance of the body of livestock as a function of the touch voltage, the degree of moisture of the hide or skin and the current path.

The report includes the following information:

▶ The internal body impedance is considered as mostly resistive and depends primarily on the current path.
▶ The impedance of a hide depends largely on humidity. In dry conditions the hide can be considered to be practically an insulator for voltages up to 100 V with impedance values in the range of tens to hundreds of kilohms.
▶ The impedance of skin depends on the voltage, frequency, duration of current flow, surface area of contact, pressure of contact, degree of moisture of the skin and temperature. The skin impedance falls as the current is increased.
▶ The impedance of the hind legs is smaller than the impedance of the forelegs. The impedance from the nose to all four legs is smaller than the impedance of forelegs to hind legs.
▶ The total body impedance (cattle) for a current path nose to the four legs is typically 35% of the total body impedance for a current path forelegs to hindlegs.

However, animals will detect relatively small voltage gradients between front and rear legs, and between conductive part potentials and Earth.

705.415.2.1

This can result in, for example, a marked reluctance for cows to enter milking parlours because of potential differences. These potential differences can arise from a number of causes. If the electricity supply is PME, at the end of a long run there is likely to be a potential between the PME earth and true Earth. In locations intended for livestock, supplementary bonding must connect all exposed-conductive-parts and extraneous-conductive-parts that can be touched by livestock. Where a metal grid is laid in the floor, it must be included within the supplementary bonding of the location (Figure 705 of BS 7671 shows an example of this; other suitable arrangements of a metal grid are not precluded). Extraneous-conductive-parts in, or on, the floor, e.g. concrete reinforcement in general or reinforcement of cellars for liquid manure, must be connected to the supplementary equipotential bonding. It is recommended that spaced floors made of prefabricated concrete elements be part of the supplementary equipotential bonding. The supplementary equipotential bonding and the metal grid, if any, must be erected so that it is durably protected against mechanical stresses and corrosion.

Note: Where a metal grid is not laid in the floor a TN-C-S supply is not recommended.

It is well known that animals have received shocks that are associated with electrical installations. It is likely that lightning strikes on overhead lines conducted to Earth via earth electrodes at the bottom of a pole produce voltage gradients that are fatal to animals because of the wide spacing of their feet. If there is concern in this respect, the location of earth electrodes should be discussed with the electricity distribution company.

5.6 Earth fault loop impedances

Where an RCD is used for earth fault protection, the following conditions are to be fulfilled for a TT installation:

411.5.3

The disconnection time must satisfy Regulation 411.3.2.2 or 411.3.2.4, and
$R_A \times I_{\Delta n} \leq 50 \text{ V}$

where:

R_A is the sum of the resistances of the earth electrode and the protective conductor connecting it to the exposed-conductive-parts (in ohms)
$I_{\Delta n}$ is the rated residual operating current of the RCD.

The above conditions would be met where the earth fault loop impedance does not exceed the values stated in Table 5.1.

Rated residual operating current, $I_{\Delta n}$ (mA)	Maximum earth fault loop impedance, Z_S (Ω)
30	1667*
100	500*
300	167

▼ **Table 5.1**
Maximum earth fault loop impedance

* The resistance of the installation earth electrode itself should be as low as practicable. A value exceeding 200 ohms may not be stable. Refer to Regulation 542.2.2.

5.7 Protection against fire

Regulation 705.422.6 requires electrical heating appliances used for the breeding and rearing of livestock to comply with BS EN 60335-2-71 and to be fixed so as to maintain an appropriate distance from livestock and combustible material, to minimise any risks of burns to livestock and of fire. For radiant heaters the clearance should be not less than 0.5 m or such other clearance as recommended by the manufacturer. For luminaires refer to Regulations 422.3.1 and 422.4.2.

Regulation 705.422.7 requires that for fire protection purposes, RCDs are to be installed with a rated residual operating current not exceeding 300 mA and must disconnect all live conductors. Where improved continuity of service is required, RCDs not protecting socket-outlets need to be of the S type or have a time delay (see Figure 5.2).

705.422.7

A note to Regulation 705.422.7 advises that the protection of final circuits by RCD required according to Regulation 411.1 is also effective for protection against fire.

Regulation 705.422.8 specifies requirements for conductors of circuits supplied at extra-low voltage in locations where a fire risk exists.

705.422.8

Rodent damage is a major cause of farm fires and this must be taken into account by the designer and installer. Cables should be installed and routed with such potential damage in mind. For example, in a livestock building, cables should be routed on the underside of the ceiling rather than in a false roof. Steel conduit provides a good degree of protection.

Additional guidance for protection against fire for this type of location can be found in section 8.2 of Guidance Note 4: *Protection Against Fire*.

GN4

▶ **Figure 5.2** Example schematic of supply arrangements (including livestock support)

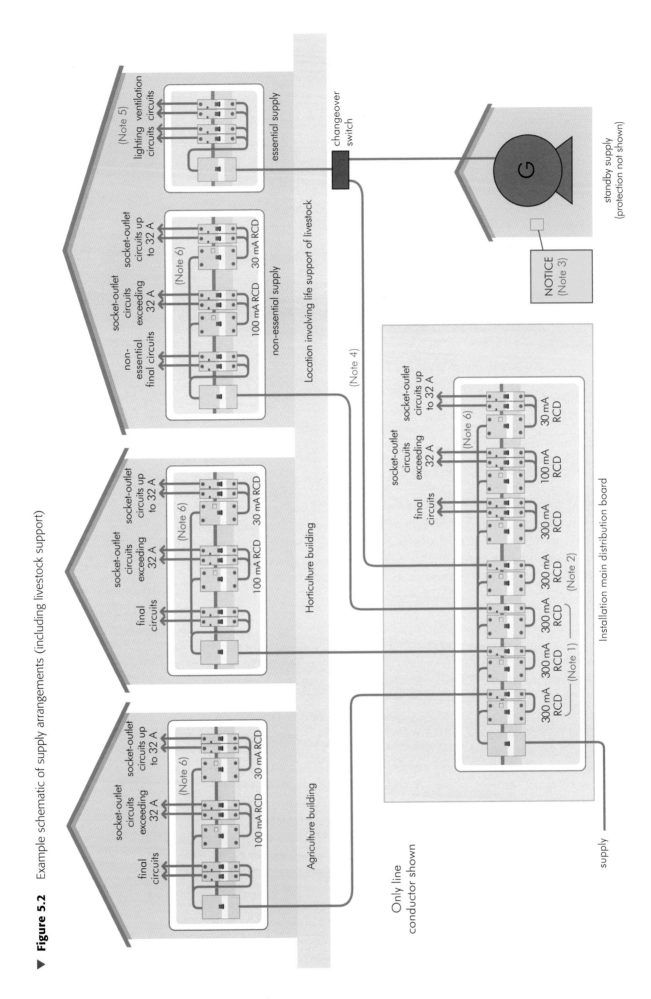

Notes to Figure 5.2:

1 All circuits other than socket-outlets must be protected by 300 mA RCDs for fire protection purposes. These will need to discriminate with final circuit RCDs where relevant.

2 For improved continuity of service, RCDs should be time delay or S type as long as they are not supplying socket-outlets.

3 A standby supply should have a notice adjacent to highlight that it should be periodically tested in accordance with manufacturer's instructions.

4 Discrimination of main circuits supplying ventilation for life support circuits.

5 Separation of lighting and ventilation circuits for life support.

6 Circuits supplying socket-outlets up to 32 A require protection by 30 mA RCDs. Circuits supplying socket-outlets exceeding 32 A require protection by 100 mA RCDs.

5.8   External influences

Electrical equipment is required to have a minimum degree of protection of IP44, when used under normal conditions. Where equipment of IP44 rating is not available, it should be placed in an enclosure complying with IP44.

705.512.2

▶ IPX4 provides protection against water splashing.
▶ IPX5 provides protection against water jets from any direction.
▶ IPX6 provides protection against powerful water jets from any direction.

See Appendix B of Guidance Note 1: *Selection & Erection* for details of the IP code and the equivalent (now superseded) drip proof, etc. symbols.

GN1

Socket-outlets should be installed in a position where they are unlikely to come into contact with combustible material. Where there are conditions of external influences >AD4, >AE3 and/or >AG1, socket-outlets must be provided with the appropriate protection. Protection may also be provided by the use of additional enclosures or by installation in building recesses.

See Appendix 5 of BS 7671:2008 for classification of external influences.

The foregoing requirements do not apply to residential locations, offices, shops and locations with similar external influences belonging to agricultural and horticultural premises where, for socket-outlets, BS 1363-2 or BS 546 applies. Where corrosive substances are present, e.g. in dairies or cattle sheds, the electrical equipment needs to be adequately protected.

Luminaires should comply with the BS EN 60598 series and be selected regarding their degree of protection against the ingress of dust, solid objects and moisture (e.g. IP54).

705.559

Luminaires marked ▽F are suitable for mounting on a normally flammable surface.

Note: The 'F' mark is to disappear (IEC 60598-1 clause 4.16) – only those luminaires which do not comply with the normal mounting requirements will need to be marked.

Luminaires marked ▽D have a limited surface temperature of the luminaire, and should have a degree of protection of IP54.

5.9 Wiring systems

705.522 Regulation 705.522 calls for wiring systems to be inaccessible to livestock or suitably protected against mechanical damage.

Where vehicles and mobile agricultural machines are operated, underground cables must be buried in the ground at a depth of at least 0.6 m with added mechanical protection. Cables in arable or cultivated ground should be buried at a depth of at least 1 m. Self-supporting suspension cables should be installed at a height of at least 6 m.

705.522.16 Regulation 705.522.16 has requirements for conduit, cable trunking and ducting systems to be protected against corrosion and impact.

5.10 Safety services

705.560.6 Where an installation includes high density livestock rearing, there may be a need to take account of continued operation of systems for the livestock (see Figure 5.2).

If the supply of food, water, air and/or lighting to the livestock cannot be ensured in the event of a power supply failure then a separate source of supply should be provided. This would include either an alternative supply or back-up supply.

For the supply of ventilation and lighting units, separate final circuits should be provided. These circuits should be limited to support just the essential equipment of the ventilation and lighting. Discrimination of the circuits supplying ventilation should be provided.

Electrically powered ventilation should include either of the following:

▶ A standby electrical source of supply that can ensure the operation of the ventilation equipment. If a standby source is used then a notice adjacent to the standby source should be provided highlighting that the standby source should be tested periodically according to the manufacturer's instructions.
▶ Monitoring of the temperature and supply voltage by one or more devices. These should provide a visual or audible signal that is located in a position that can be readily observed by the user, and the device(s) should operate independently from the normal supply.

Conducting locations with restricted movement

6

6.1 Introduction

The section on conducting locations with restricted movement (titled 'Restrictive conductive locations' in the 16th Edition) is retained. The only notable change in the 17th Edition is the addition of PELV to the list of suitable protective measures, subject to the necessary bonding being in place.

Section 706

6.2 Scope

A *conducting location with restricted movement* is constructed mainly of metallic or other conductive surrounding parts, and within it movement is restricted. It is likely that a person in such a location will be in good contact with conductive surroundings and escape will be difficult in the event of an electric shock. The regulations apply to the fixed conducting location and to the supplies for mobile and hand-held equipment for use in such locations.

Part 2

706.1

The particular requirements of Section 706 do not apply to locations which allow persons freedom of bodily movement, that is, to enter, work and leave the location without physical constraint. The types of location that are being considered include boiler shells, cable gantries, small tunnels, metal sewers, etc.

6.3 The risks

In many ordinary locations, there is usually limited access to earthed metal. As a result, the likelihood of receiving a shock current of sufficient magnitude to have harmful physiological effects is low. This is not so with a conducting location.

In a conducting location where bodily movement is limited, there is little opportunity to move away from the shock. Contact resistance is low due to high contact areas and perspiration, so that body currents are high and the risk of ventricular fibrillation is also high. There are other effects of electric shock that are also relevant in such locations. Muscles used to breathe can be constrained by body currents, and shocks to the head can paralyse the breathing function.

6.4 Protection against electric shock

706.410.3.5   The protective measures of obstacles and placing out of reach are not permitted in these locations.

706.410.3.10   For supplies to equipment, the following protective measures are permissible:

Hand-held tools or items of mobile equipment
a   electrical separation, only one item of equipment being connected to a secondary winding of the transformer (which may have two or more such windings), or

b   SELV.

Handlamps
a   SELV

Fixed equipment
a   automatic disconnection of supply with supplementary equipotential bonding connecting exposed-conductive-parts of fixed equipment and the conductive parts of the location, or

b   use of Class II equipment or equipment having equivalent insulation, provided the supply circuits have additional protection by the use of an RCD with a rated residual operating current not exceeding 30 mA having the characteristics specified

415.1.1   in Regulation 415.1.1, or

c   electrical separation, only one item of equipment being connected to a secondary winding of the transformer, or

d   SELV or PELV.

For PELV, equipotential bonding must be provided between all exposed-conductive-parts, extraneous-conductive-parts inside the location and the connection of the PELV system to Earth.

706.413   For electrical separation, in addition to the requirements of Section 413, the unearthed source should have simple separation and should be situated outside the conducting location with restricted movement, unless the source is part of the fixed electrical installation of the location.

706.414   For SELV or PELV, the general requirements of Section 414 apply, except that, whatever the nominal voltage, basic protection must be provided. In addition, the source of SELV or PELV must be situated outside the conducting location with restricted movement, unless it is part of the fixed installation.

Electrical installations in caravan/camping parks, caravans and motor caravans

7

7.1 Introduction

The 16th Edition included two separate divisions within the same section. The first of these addressed caravans and motor caravans, the second addressed caravan parks.

The 17th Edition has two separately numbered sections, namely Section 708 concerning the electrical installations in caravan parks, etc., and Section 721 concerning caravans and motor caravans.

Section 708
Section 721

This chapter provides guidance on applying the particular requirements of both Sections 708 and 721 of BS 7671:2008.

7.2 Scope of Sections 708 and 721

7.2.1 Section 708
The particular requirements of this section apply to that portion of the electrical installation in caravan/camping parks and similar locations providing facilities for supplying leisure accommodation vehicles (including caravans) or tents.

The requirements do not apply to the internal electrical installations of leisure accommodation vehicles or mobile or transportable units.

7.2.2 Section 721
The particular requirements of this section apply to the electrical installations of caravans and motor caravans at nominal voltages not exceeding 230/440 V a.c. or 48 V d.c., except for 12 V d.c. – see exclusions below.

The requirements do not apply to the following:

▶ electrical circuits and equipment intended for the use of the caravan for habitation purposes
▶ electrical circuits and equipment for automotive purposes
▶ 12 V d.c. installations, which are covered by BS EN 1648-1 and BS EN 1648-2
▶ electrical installations of mobile homes or residential park homes, to which the general requirements apply
▶ transportable units (see Chapter 14 of this Guidance Note).

For the purposes of Section 721, caravans and motor caravans are referred to as 'caravans'. The particular requirements of some other sections of Part 7 may also apply to electrical installations in caravans, e.g. Section 701 (showers).

7.3 The risks

The risks specifically associated with installations in caravan parks, caravans and motor caravans arise from:

i   open circuit faults of the PEN conductor of PME supplies raising the potential to true Earth of all metalwork (including that of caravans if connected) to dangerous levels

ii   incorrect polarity at the pitch supply point

iii   inability to establish an equipotential zone external to the vehicle

iv   possible loss of earthing due to long supply cable runs, connecting devices exposed to weather and flexible cord connections liable to mechanical damage

721.522.7.1   v   vibration while the vehicle is moving, causing faults within the caravan installation.

Particular requirements to reduce the above risks include:

708.553.1.14   i   prohibition of the connection of exposed- and extraneous-conductive-parts of a caravan or motor caravan to a PME terminal

721.411.1   ii   additional protection by 30 mA RCDs in both the vehicle and the park installation

708.553.1.13   iii   double-pole isolating switch and final circuit circuit-breakers in the caravan or motor
721.537.2.1.1   caravan

721.43.1   iv   internal wiring of the caravan or motor caravan by flexible or stranded cables of
721.524.1   cross-sectional area 1.5 mm² or greater; additional cable supports; segregation of
721.522.8.1.3   low voltage and extra-low voltage circuits.
721.528.1

7.4 Legislation and standards

Regulation 9(4) of the Electricity Safety, Quality and Continuity Regulations 2002 prohibits the connection of the supply neutral of a PME supply to any metalwork in a caravan or boat.

Caravan parks (sites) in the United Kingdom are subject to the provisions of the Caravan Sites and Control of Development Act 1960. This empowers local authorities to issue licences and to impose conditions, generally in accordance with model standards. For residential parks there are the *Model Standards 2008 for Caravan Sites in England* and *Model Standards 2008 for Caravan Sites in Wales*. For holiday sites there are the *Model Standards 1989: Holiday Caravan Sites*. For touring sites there are the *Model Standards for Touring Caravan Sites*. These model standards include requirements for the electrical installation in the caravan park to be installed and maintained to the requirements of the current edition of the *IEE Wiring Regulations* (BS 7671).

The 1960 Act also empowers certain 'exempted bodies' such as the Caravan Club and the Camping and Caravanning Club to issue certificates in respect of parks for use by their own members.

While there is no legislation specific to electrical installations in caravans, BS EN 1645-1 *Leisure accommodation vehicles – caravans* includes a section on electrical installations. This requires that low voltage electrical installations are to comply with international standard IEC 60364-7-708 *Low voltage electrical installations – Part 7-708:*

Requirements for special installations or locations – Caravan parks, camping parks and similar locations. These requirements are included in Section 708 of BS 7671:2008.

Extra-low voltage installations using 12 V d.c. should comply with BS EN 1648-1 *Leisure accommodation vehicles – 12 V direct current extra low voltage electrical installations – Part 1: Caravans.*

7.5   Electrical installations in caravan parks

7.5.1   Supply systems

The supply system to the permanent buildings of a caravan park can be TN-C-S, TN-S or TT. Consideration should be given to the earthing and bonding of amenity buildings such as toilet and shower blocks. The supply system to the caravans is limited to TN-S or TT, as the Electricity Safety, Quality and Continuity Regulations 2002 prohibit the use of a TN-C-S system to supply a caravan or similar construction. If the caravan supply is derived from a permanent building that is supplied by a TN-C-S system then the caravan supply will have to be part of a TT system having a separate connection to Earth independent from the TN-C-S earthing.

708.411.4

The separation of the earthing should preferably be effected at the main distribution board (see Figure 7.1). This enables the exposed-conductive-parts connected to each system to be more readily identified and inspected periodically. An earth electrode for the TT system should be provided nearby and located so that the resistance areas of the TN-C-S supply earthing and earth electrode do not overlap (refer to section 7.5.5).

542.1.8

Alternatively, the separation of the earthing can be made at the caravan pitch supply points. In this instance, earth electrodes will be required at these points. A cable supplying a separate earthing system is to be earthed only at the installation containing the associated protective device (see Figure 7.2).

For automatic disconnection of supply, final circuits not exceeding 32 A require maximum disconnection times of 0.4 s for TN and 0.2 s for TT systems. Final circuits exceeding 32 A and distribution circuits require maximum disconnection times of 5 s for TN and 1 s for TT systems. For TT supplies an RCD of 100 mA or more will be required which will need to be a delay or 'S' type to discriminate with the individual 30 mA RCDs required for the pitch outlets. If an RCD is used for a TN-S system then discrimination will again be required.

411.3.2.2
411.3.2.3
411.3.2.4
314.1(i)

7.5.2   Overhead and underground distribution

BS 7671 states a preference for distribution by underground cable. Underground cables should be buried at a depth of at least 0.6 m and should be routed outside any caravan pitch or away from areas where tent poles or ground anchors could be used, unless additional mechanical protection is provided. Overhead distribution systems are allowed provided that all conductors are insulated and so erected as to be unlikely to be damaged by vehicle movement. They are required to be not less than 6 m above ground level in areas subject to vehicle movement and 3.5 m in other areas. Suitable warning notices should be displayed at the entrance to the site and on supports for the overhead line and, where appropriate, attention drawn to the danger of masts of yachts or dinghies contacting the overhead line.

708.521.1

708.410.3.5

▶ **Figure 7.1** Typical site distribution for a PME supply, separation from PME earth at main distribution board

pitch socket-outlets

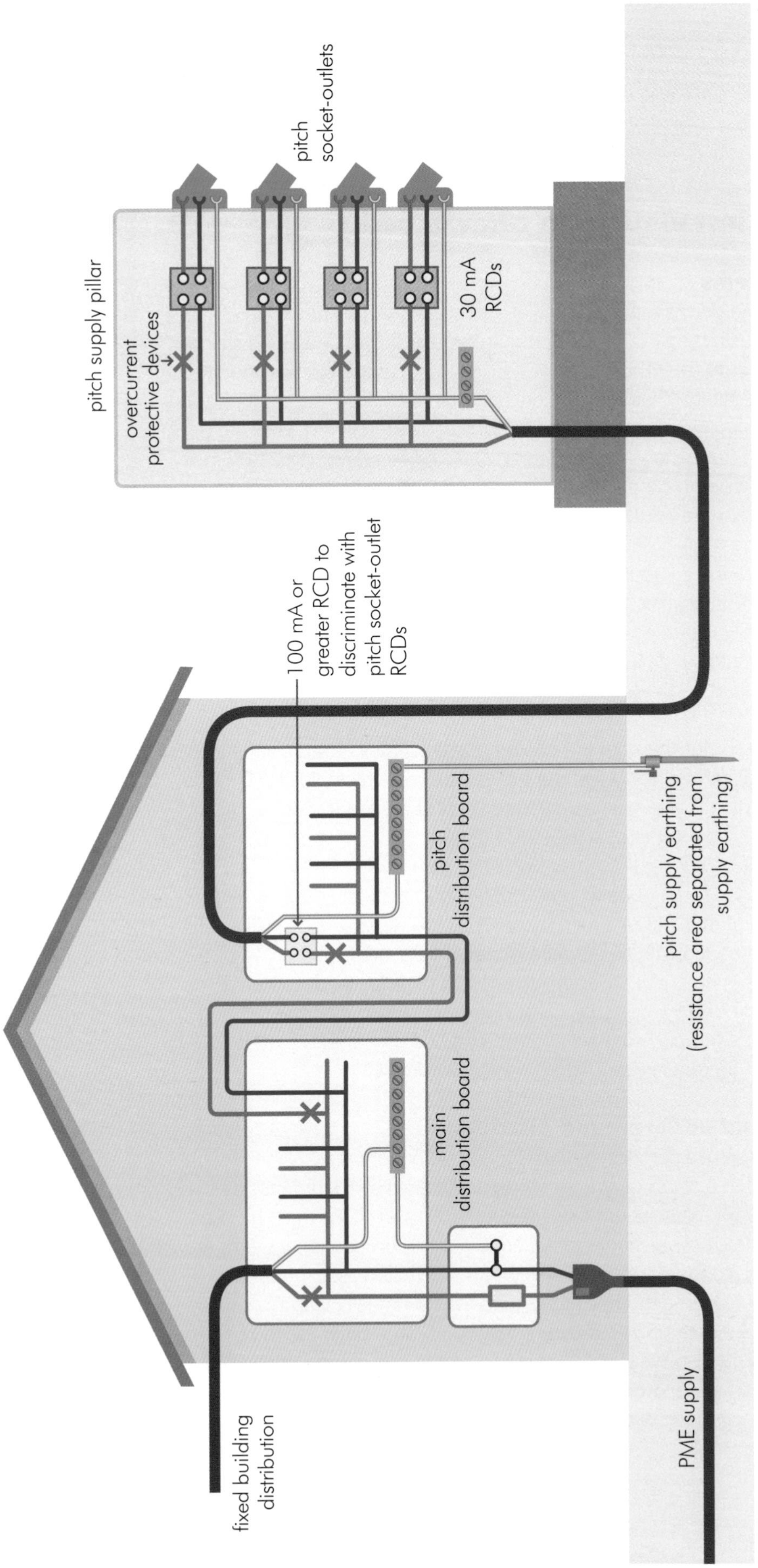

pitch supply pillar

overcurrent protective devices

30 mA RCDs

100 mA or greater RCD to discriminate with pitch socket-outlet RCDs

pitch distribution board

main distribution board

fixed building distribution

pitch supply earthing (resistance area separated from supply earthing)

PME supply

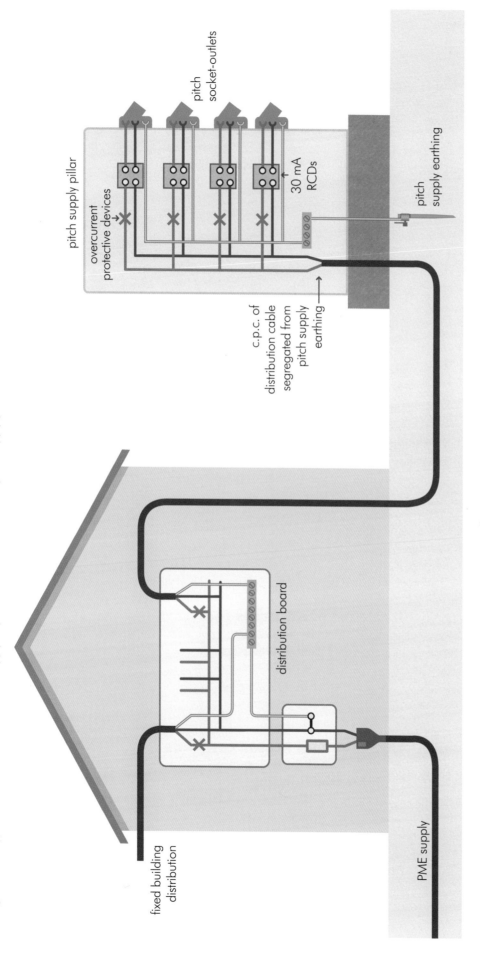

▶ **Figure 7.2** Typical site distribution for a PME supply, separation from PME earth at pitch supply point

7.5.3   Caravan pitches

Dimensions of pitches are provided within the model standards issued under the Caravan Sites and Control of Development Act 1960, Section 5. Dimensions of pitch relate to the type of site as follows:

▶ **For a holiday site.** There should be a 3 m wide area kept clear within the inside of all boundaries. For units made of aluminium or similar fire performance properties, there should be a minimum of 5 m between units (3.5 m at the corners). For units made of plywood or similiar, there should be a minimum of 6 m. For a mixture of types a minimum of 6 m. For a mixture of residential homes and holiday caravans a minimum of 6 m.

▶ **For a touring site.** There should be a 6 m minimum distance between units in separate family occupation and 3 m minimum between units in any circumstances.

▶ **For a residential site.** There should be a 6 m minimum distance between caravans. If a caravan has been fitted with cladding from Class 1 fire rated materials then the minimum distance is 5.25 m.

When designing pitches, the requirements of the above legislation should be confirmed with the relevant local authority.

7.5.4   Caravan pitch socket-outlets

708.553.1.8   Caravan pitch socket-outlets are required to comply with BS EN 60309-2 and to have
708.553.1.10   a degree of protection of at least IP44. The current rating is to be not less than 16 A but may be greater if required. At least one socket-outlet should be provided for each caravan pitch. Where socket-outlets are grouped in pitch supply equipment, there
708.530.3   should be one socket for each pitch limited to a group of four. To be compatible with the caravan connecting cable, sockets should be two-pole with earthing contact having key position 6 h.

707.553.1.12   Each socket-outlet must be protected individually by an overcurrent device, which may
708.553.1.13   be a fuse but is more usually a circuit-breaker, and individually by a 30 mA RCD having the characteristics specified in Regulation 415.1.1. The RCD should disconnect all live conductors including the neutral. See Figures 7.1 and 7.2.

7.5.5   Separation of electrodes

Figure 7.3 indicates that effective separation of resistance areas of earth electrodes is achieved if the distance between the electrodes exceeds 10 m.

GN3   Guidance Note 3: *Inspection & Testing* describes a test method for measurement of earth electrode resistance.

7.6   Caravans and motor caravans

7.6.1   Equipotential bonding

721.411.3.1.2   Structural metallic parts which are accessible from within the caravan are required to be connected through main protective bonding conductors to the main earthing terminal within the caravan.

7.6.2   Provision of RCDs

721.411.1   Where protection by automatic disconnection of supply is used (Section 411), an RCD complying with BS EN 61008-1 or BS EN 61009-1 interrupting all live conductors

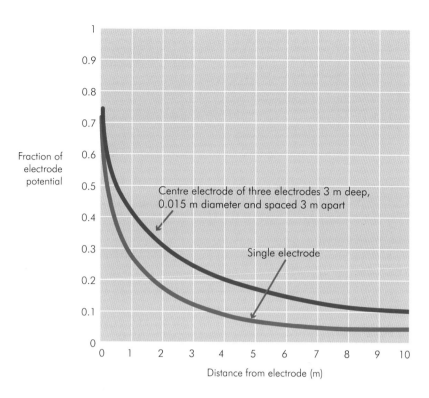

is required to be provided having the characteristics specified in Regulation 415.1.1 (30 mA), and the wiring system must include a circuit protective conductor connected to:

i   the protective contact of the inlet, and
ii the exposed-conductive-parts of the electrical equipment, and
iii the protective contacts of the socket-outlets.

7.6.3   Protection against overcurrent

Each final circuit must be protected by an overcurrent protective device which disconnects all the live conductors of the circuit.

721.43.1

7.6.4   Selection and erection of equipment

More than one electrically independent installation is permitted provided each one is supplied via a separate connecting device and they are segregated in accordance with the requirements of Regulation 528.1. This includes requirements for insulating cables or conductors to the highest voltage present, or installing the independent circuit cables in separate conduits, trunking or ducting compartments.

721.510.3

528.1

This enables 12 V d.c. battery supplied circuits for interior lighting in accordance with BS EN 1648 *Leisure accommodation vehicles 12 V direct current extra low voltage installations, Part 1 – caravans, Part 2 – motor caravans* and for the road lighting circuits of the caravan or motor caravan in accordance with the correct amendment of the Road Vehicle Lighting Regulations 1989.

7.6.5   Switchgear and controlgear

The installation to the caravan should have a main disconnector which will disconnect all the live conductors. This should be placed in a suitable position ready for operation within the caravan to isolate the supply. When a caravan only has one final circuit then the isolation can be afforded by the overcurrent protective device as long as it fulfils the requirements for isolation.

721.537.2

Figure 721
An indelible notice in the appropriate language(s) must be permanently fixed near the main isolation point inside the caravan to provide the user with instructions on connecting and disconnecting the supply (refer to Figure 721 of BS 7671).

721.55.1 The inlet to the caravan must be an appropriate inlet complying with BS EN 60309. This should be installed not more than 1.8 m above ground level, in a readily accessible position, have a minimum degree of protection of IP44, and should not protrude significantly beyond the body of the caravan.

7.6.6 The connecting flexible cable

721.55.2.6 The means of connection of the caravan to the pitch socket-outlet should be provided with the caravan. This must have a plug at one end complying with BS EN 60309-2, a flexible cord or cable with a length of 25 m (±2 m) and a connector if needed that is compatible with the appropriate appliance inlet. The cable should be to the harmonized code H05RN-F (BS 7919) or equivalent, include a protective conductor, have cores coloured as required by Table 51 of BS 7671 and have a cross-sectional area as shown in Table 7.1.

▼ **Table 7.1**
Minimum cross-sectional areas of flexible cords and cables for caravan connection (Table 721 of BS 7671:2008)

Rated current (A)	Minimum cross-sectional area (mm²)
16	2.5
25	4
32	6
63	16
100	35

7.7 Tents

BS 7671 does not provide specific requirements for electrical installations in tents; however, electrical installations in tents are common in other countries, especially in permanently erected tents offered by holiday firms, the supply being derived from the park installation. For trailer tents, mains-supply units should be used which incorporate the equipment specified for a caravan, namely a main double-pole isolating switch and a main double-pole RCD (which may be combined), one or more double-pole circuit-breakers and socket-outlets as required. Double-pole circuit-breaker means having overcurrent detection and switching in both poles and not a single-pole circuit-breaker with a switched neutral. The connecting cable should be securely clamped to the main supply unit and directly connected to it without the use of a plug and inlet. Luminaires and appliances should be of Class II construction.

Marinas and similar locations 8

8.1 Introduction

The 16th Edition did not include requirements for marinas, but guidance was given in the previous edition of this Guidance Note. However, new Section 709 concerning electrical installations of marinas has been introduced into the 17th Edition.

Section 709

Section 709 specifies particular requirements for this type of electrical installation relating to assessment of general characteristics, protection against electric shock and selection and erection of equipment.

For electrical installations of pleasure craft, reference should be made to BS EN 60092-507 *Electrical installations in ships – Pleasure craft*, while for houseboats the general requirements of BS 7671:2008 apply.

8.2 Scope

The particular requirements of Section 709 are applicable only to circuits intended to supply pleasure craft or houseboats in marinas and similar locations. (In this section 'marina' means 'marina and similar locations'.)

709.1

The particular requirements do not apply to the supply to houseboats if they are supplied directly from the public network, or to the internal electrical installations of pleasure craft or houseboats.

For the remainder of the electrical installation of marinas and similar locations the general requirements of the Regulations apply, together with any relevant particular requirements of Part 7.

8.3 The risks

The environment of a marina or yachting harbour is harsh for electrical equipment. The water, salt, and movement of structures accelerate deterioration of the installation. The presence of salt water, dissimilar metals and a potential for leakage currents increases the rate of corrosion. There are also increased electric shock risks associated with a wet environment, by reduction in body resistance and contact with Earth potential.

709.512.2

Site investigations should be carried out at an early stage to determine likely maximum wave heights. This is of particular importance in exposed coastal sites. Where marinas have breakwater type pontoons, it is likely that under certain conditions waves will pass over the structure.

The risks specifically associated with craft supplied from marinas include:

i   open circuit faults of the PEN conductor of PME supplies raising the potential to true Earth of all metalwork (including that of the craft, if connected) to dangerous levels

ii   inability to establish an equipotential zone external to the craft

iii possible loss of earthing due to long supply cable runs, connecting devices exposed to weather and flexible cord connections liable to mechanical damage.

Particular requirements to reduce the above risks include:

i   prohibition of the connection of exposed- and extraneous-conductive-parts of the craft to a PME terminal

ii   additional protection by 30 mA RCDs in both the craft and the marina installation.

8.4   General requirements

Electrical power installations located at marinas should be installed and the equipment so selected as to minimise the risk of electric shock, fire and explosion. In the design and construction of such works, particular regard should be given to the risk of increased corrosion, movement of structures, mechanical damage, presence of flammable fuel and vapour and the physiological effects of electric shock being increased by a reduction in body resistance and contact of the body with earth potential.

Owing to the harsh working environment of marina installations and potential for abuse and accidental damage by users, particular attention should also be paid to the maintenance and periodic inspection reporting of installations and the general requirements of the Regulations.

8.5   Supplies

Where the supply system is protective multiple earthed (PME), Regulation 9(4) of the Electricity Safety, Quality and Continuity Regulations 2002 prohibits the connection of the neutral to the metalwork of any caravan or boat. While the PME supply may be fed to permanent buildings in the marina, supplies to boats (pleasure craft) must have a separate earth system. A TT system having a separate connection with Earth, independent of the PME earthing system (see Figure 8.2a/b and Figures 7.1 and 7.2), will meet this requirement. Alternatively, protection by electrical separation can be adopted; see section 8.7 and Figure 8.2c.

This avoids the risks arising from a loss of continuity of the supply PEN conductor.

The separation of the TT earthing system should be effected at the main distribution board, where the exposed-conductive-parts connected to each system can be more readily identified and inspected periodically. A main earth electrode for the TT system needs to be provided nearby, with no overlap of resistance area with any earthing 541.2 associated with the PME supply. (See also section 7.5.5 of Chapter 7.)

TN-S supplies may be made available both to permanent shore installations and to pleasure craft.

The nominal supply voltage of the installation to pleasure craft should not exceed 230 V single-phase or 400 V three-phase.

Some locations may have dry dock areas where construction and maintenance activities can be carried out. The guidance on supplies to these locations can be summarised as follows:

▶ For dry dock areas that form part of a formal workshop type environment under responsible control where construction and maintenance on vessels are carried out, reduced low voltage (110 V centre-tap earthed) and extra-low voltage fixed supplies should be provided. In addition, portable generators could be used that would require the appropriate earthing and protection.

▶ British Waterways have a number of dry dock locations accessible by vessel owners where maintenance activities can be carried out. Here, typical vessel distribution pillar fixed supplies with the appropriate overcurrent and RCD protection would be provided. The vessel owner would connect to this supply and use double insulated electrical equipment connected via the vessel socket-outlets.

▶ British Waterways also have a number of dry dock locations that do not have any fixed installation electrical supply. The supply in this type of location would be via a small generator. This would require the appropriate earthing and protection to be provided.

Note: Persons involved in designing and erecting electrical installations for dry dock facilities should contact the Health and Safety Executive and British Waterways for additional guidance.

8.6 Protection against electric shock

The protective measures of obstacles and placing out of reach are not permitted. Also, the protective measures of non-conducting location and earth-free local equipotential bonding are not permitted.

709.410.3

In the UK for a TN system, the final circuits for the supply to pleasure craft or houseboats shall not include a PEN conductor.

709.411.4

Note: In the UK the ESQCR prohibit the use of a TN-C-S system for the supply to a boat or similar construction.

Only permanent onshore buildings may use the electricity distributor's PME earthing terminal. For the boat mooring area of the marina this is not permissible, and entirely separate earthing arrangements must be provided. This is generally achieved by the use of a suitably rated RCD complying with BS EN 61008 with driven earth rods or mats providing a TT system for that part of the installation.

Marina installations are often of sufficient size to warrant the provision of an 11 kV/415 V transformer substation. In these, and sometimes in other, circumstances the electricity distributor may be willing to provide a TN-S supply, which is much more suitable for such installations. If the transformer belongs to the marina, a TN-S system should be installed.

8.7 Isolating transformers

Supplies to craft may be provided from any supply system through isolating transformers. This method has the advantage of reducing electrolytic corrosion and can be used with TN-S and TN-C-S (PME) supplies.

Isolating transformers must comply with BS EN 60742 *Isolating transformers and safety isolating transformers* or the BS EN 61558 series. See Figure 8.2c for typical wiring arrangement with onshore mounted isolating transformers.

Connection of the protective conductor of the shore supply must not be made to the bonding of the pleasure craft. However, the following items must be effectively and reliably connected to a bonding conductor – which, in turn, must be connected to one of the secondary winding terminals of the isolating transformer:

▶ Metal parts of the pleasure craft which are in electrical contact with water. If the type of construction does not ensure continuity, then more than one connection point may be required
▶ The protective contact of each socket-outlet
▶ The exposed-conductive-parts of electrical equipment.

Only one craft (socket-outlet) must be connected to each secondary winding of an isolating transformer.

Note: The isolating transformer isolates the craft installation from the shore, allowing supplies to be taken from multiple earthed networks, and provides some protection against electrolytic corrosion. It does not provide basic or fault protection; the craft earth is connected to one pole of the secondary isolating transformer.

8.8 Operational conditions and environmental factors

709.512.2.1.1
709.512.2.1.2

Electrical equipment to be installed on or above jetties, wharves, piers or pontoons must be selected according to the external influences which may be present. Regarding presence of solid foreign bodies, a minimum degree of protection of IP3X is required, and for presence of water the following applies.

External influence	Minimum index of protection
Presence of water splashes	IPX4
Presence of water jets	IPX5
Presence of waves of water	IPX6

709.512.2.1.3 In the marina environment, particularly at jetties, pontoons etc., consideration must also be given to the possible presence of corrosive or polluting substances.

709.512.2.1.4 Equipment should be located to avoid any foreseeable impact, be provided with local or general mechanical protection and have a degree of protection for external mechanical impact IK08 (see BS EN 62262).

Specific guidance for distribution boards and socket-outlets is given in section 8.10.

8.9   Wiring systems

The following wiring systems should **not** be used above a jetty, wharf, pier or pontoon: 709.521.1.5

▶ Cables suspended from or incorporating a support wire
▶ Non-sheathed cables in conduit, trunking, etc.
▶ Cables with aluminium conductors
▶ Mineral insulated cables.

Conduit and ducting installations should have suitable apertures or holes and be fixed at an angle sloping away from the horizontal, sufficient to allow for drainage of moisture.   709.521.1.6

Cables should be selected and installed so that mechanical damage due to tidal and other movement of craft and other floating structures is prevented. To clarify this requirement, cables should be installed in such a manner that they are protected from damage due to:

▶ displacement by movement of craft or other structures
▶ friction, tension or crushing
▶ exposure to adverse temperatures.

See Figure 8.1 for a typical wiring arrangement for offshore pontoons.

At locations where cables are subject to flexing, e.g. bridge ramps, between movable jetties and pontoons, flexible cables should be used, such as:

▶ cross-linked insulated flexible cables harmonized type H07RN-F, H07BN4-F or H07RN8-F (insulated and sheathed), e.g. cables to Tables 14, 15, 16, 17 and 20 of BS 7919:2001
▶ thermosetting insulated flexible cables harmonized type H07Z-K, e.g. cables to BS 7211 Table 3b within flexible wiring systems.

Notes
1   Cables should be installed in locations where they are protected from physical damage and, wherever practicable, out of water.
2   Many cable types including PVC insulated and sheathed cables are not suitable for continuous immersion in water. The suitability of the cable types should be checked with the manufacturers. Floating pontoons are usually manufactured with a service void in them, enclosed and accessible from above, to accommodate cables and water piping.
3   Fixed cables installed permanently under water at a depth of more than 4 m will normally need to be metal sheathed, e.g. lead. Alternatively, a cable that has been specially designed and manufactured for such locations could be used (refer to cable manufacturer for further guidance).
    Fixed cables not permanently immersed or at a depth of less than 4 m should be armoured and incorporate extruded MDPE (medium density polyethylene) outer sheath. Note that cables used in this location will not have the same life expectancy as similar cables used on dry land.
4   Due to the possibility of corrosion, the galvanised steel armouring of cables must not be used wholly or in part as a circuit protective conductor (cpc) on the floating section of marinas. A separate protective conductor should be used which, when

8

▶ **Figure 8.1** Typical wiring arrangement from shore to pontoon

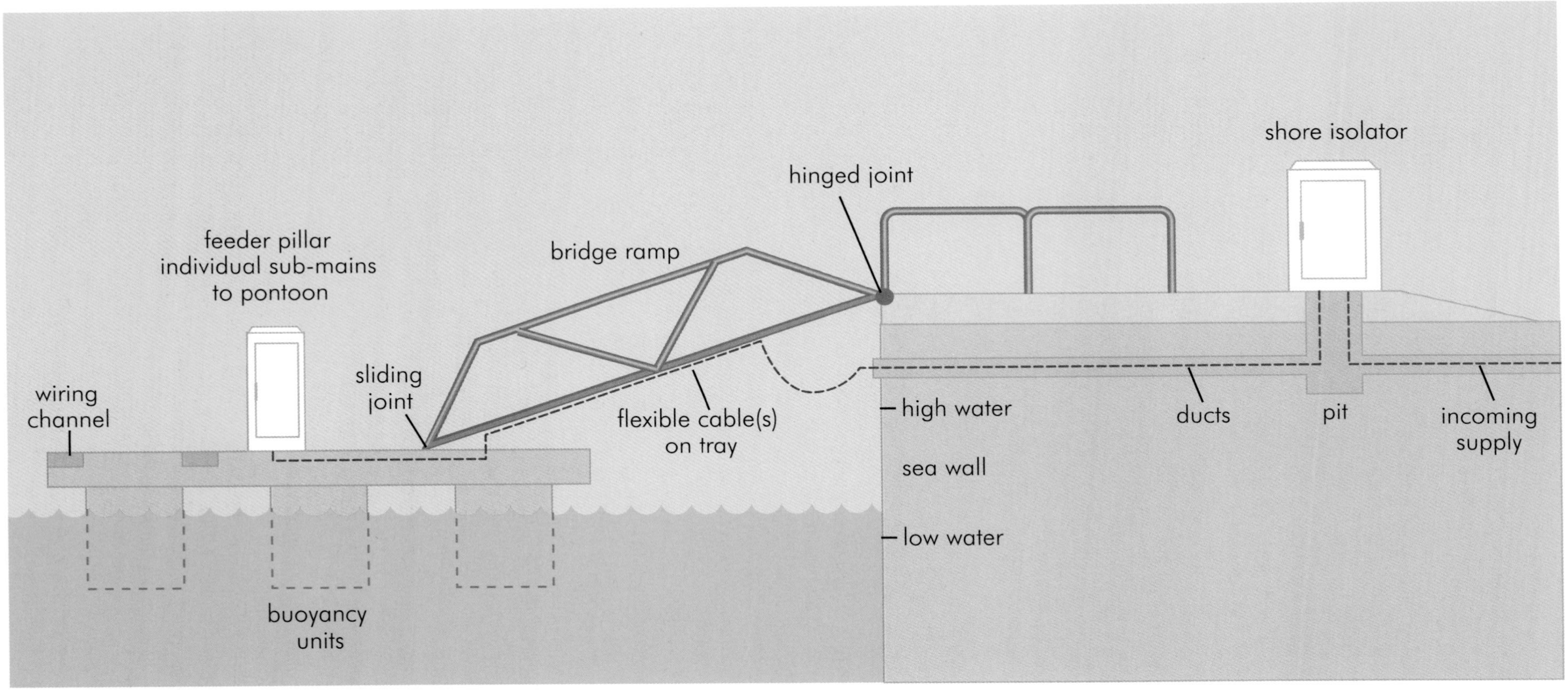

Notes:

1 Where the particular feeder pillars are in external locations they should be constructed of glass reinforced plastic (GRP), or have GRP housings. GRP is preferred to galvanised steel for protection against corrosion in such environments.

2 In order to counteract condensation within feeder pillar enclosures, low wattage 'anti-condensation' heaters should be installed.

3 All feeder pillar and distribution board doors should be fitted with locks to prevent unauthorised access, and have intermediate barriers to protect against accidental contact with live parts when the doors are open. The barriers should provide a degree of protection of at least IP2X or IPXXB.

sized in accordance with Regulation 543.1.2, can be common to several circuits if necessary. The armour must still, however, be connected to protective earth.   543.1.2

5 Protective bonding connections must be single-core PVC insulated to BS 6004 (HAR reference – H07V-R and H07Z-R), or BS 6007 (flexible type), or with an oversheath or further mechanical protection as applicable to the particular location.

6 Conductor colour coding should be in accordance with the requirements of BS 7671 Table 51. Terminations should be protected against corrosion either by the   Table 51 selection of suitable materials or covering with grease or water-resistant mastic or paint.

7 Care should be exercised when installing cables to prevent damage from abrasion due to movement between pontoon sections, etc. Cables must be adequately fixed, protected and supported, and, if necessary, cable types suitable for the flexing movement must be used.

8 Where cables are installed at onshore locations, due consideration should be given to the routing, depth of lay and protection, especially where heavy traffic and point loads are experienced. Cables should normally be laid above the water table, or cable types suitable for continual immersion used. It is not usually practicable for buried cable duct systems to be made totally watertight. The watertight termination of ducts into drawpits and cable trenches below switchboards is also difficult to achieve.

8.10 Distribution boards, feeder pillars and socket-outlets

Distribution boards and feeder pillars mounted outdoors should meet the degree of protection IP44 as a minimum. This will be adequate in sheltered waterways, but the IP code must be selected with reference to the degree of protection necessary for the particular location. The enclosure should be corrosion resistant and give protection against mechanical damage and ingress of dust and sand etc. Distribution boards and feeder pillars supplying marina berths should be sited in the immediate vicinity of berths. When distribution boards and feeder pillars and their associated socket-outlets   709.553.1.13 are mounted on floating installations or jetties, they should be fixed above the walkway and preferably not less than 1 m above the highest water level. This height may be reduced to 300 mm if appropriate additional measures are taken to protect against the effects of splashing (IPX4), but care should be taken to avoid creating a low-level obstacle which may cause risk of tripping on the walkway. When mounted on fixed jetties they should be mounted not less than 1 m above the highest water level.

Note: While there is a good argument for socket-outlets on feeder pillars and bollards to be mounted at a high level, they may be at risk from damage from the bows of boats which can accidentally overshoot the walkways during berthing. A lower mounting level of 300 mm minimum above the walkway can reduce this risk, but care should be taken to avoid creating a low level obstacle, which may trip the unwary.

Socket-outlets should be mounted as close as possible to the berth to be supplied   709.553.1.9 and should be installed in a distribution board or separate enclosures. A maximum of   709.553.1.10 four socket-outlets may be grouped together in one enclosure. This is to minimise the hazard of long trailing flexes. A socket-outlet, either single-phase or polyphase, is only to supply a single pleasure craft or houseboat.

Socket-outlets should be in accordance with BS EN 60309-2 and each outlet should   709.553.1.8 be connected to the circuit protective conductor except where an onshore isolating transformer is used. For isolating transformers, the socket-outlet protective conductor is made to one end of the secondary winding of the isolating transformer and remains separated from the shore earthing (see Figure 8.2c).

709.553.1.12 Generally, socket-outlets with a rating of 16 A should be provided and should have the following characteristics, irrespective of the measure of protection against electric shock:

Single-phase socket-outlets
Rated voltage: 230 V (colour blue)
Rated current: 16 A
Key position: 6 h
Number of poles: 2 plus protective conductor
Construction: IP44 (minimum)

Three-phase socket-outlets
Rated voltage: 400 V (colour red)
Rated current: 16 A
Key position: 6 h
Number of poles: 4 plus protective conductor
Construction: IP44 (minimum)

Note: Whilst 2P + E and 4P + E plugs and sockets are generally used, other configurations may be necessary as in the case of special security circuits indicating unauthorised use of particular socket-outlets on remote monitoring systems.

709.553.1.12 Where the pleasure craft demand is likely to exceed 16 A, provision should be made for outlets of suitable rating.

709.531 For automatic disconnection of supply, socket-outlets to supply pleasure craft and final circuits intended for fixed supplies to houseboats are to be protected individually by a 30 mA RCD having the characteristics specified in Regulation 415.1.1. The devices will need to disconnect all poles including the neutral (see Figures 8.2a–c).

709.533 Socket-outlets and fixed connection supplies to houseboats are to be protected by individual overcurrent protective devices, which should comply with the requirements of Chapter 43 of BS 7671:2008 (see Figures 8.2a–c).

709.537.2.1.1 For a distribution cabinet there should be at least one means of isolation that will disconnect all live conductors including the neutral. If the device is used to isolate socket-outlets then it should isolate not more than four of these.

Preferably, socket-outlets or groups of single-phase socket-outlets intended for use on the same walkway or jetty should be connected to the same phase. However, individual sockets connected to separate phases of a supply should be located so that **514.10** they cannot be reached simultaneously. If sockets on separate phases are grouped together on a pillar then a notice warning of the maximum voltage that exists between accessible parts should be provided.

Socket-outlets should include an interlock to prevent the insertion or removal of a plug while under load.

A notice of durable material giving instructions for connection of a pleasure craft to the marina supply is recommended to be placed where practicable adjacent to each group of socket-outlets, bearing indelible, weatherproofed and easily legible characters.

▼ **Figure 8.2**   General arrangements for electricity supply to pleasure craft

a Connection to mains supply with single-phase socket-outlet

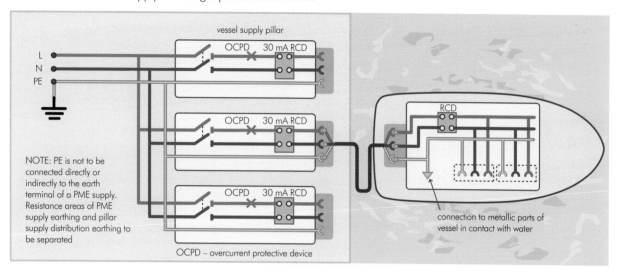

b Connection to mains supply with three-phase socket-outlet

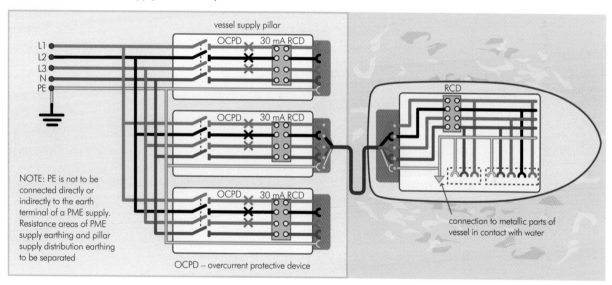

c Onshore mounted isolating transformers (hull and metal parts of craft bonded)

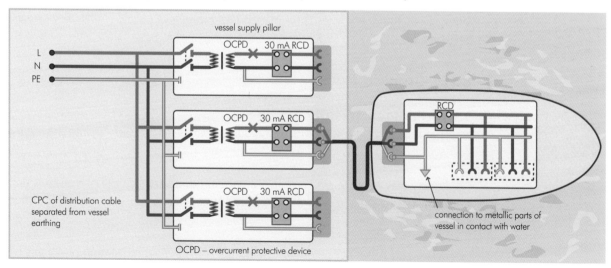

▼ **Figure 8.3**
Instruction notice for
connection of supply
(Figure 709.3 of
BS 7671:2008)

INSTRUCTIONS FOR ELECTRICITY SUPPLY

BERTHING INSTRUCTIONS FOR CONNECTION TO SHORE SUPPLY

This marina provides power for use on your pleasure craft with a direct connection to the shore supply which is connected to earth. Unless you have an isolating transformer fitted on board to isolate the electrical system on your craft from the shore supply system, corrosion through electrolysis could damage your craft or surrounding craft.

ON ARRIVAL

1 Ensure the supply is switched off and disconnect all current-using equipment on the craft, before inserting the craft plug. Connect the flexible cable **firstly** at the pleasure-craft inlet socket and **then** at the marina socket-outlet.

2 The supply at this berth is *V, *Hz. The socket-outlet will accommodate a standard marina plug colour * (technically described as BS EN 60309 2, position 6 h).

3 For safety reasons, your craft must not be connected to any other socket-outlet than that allocated to you and the internal wiring on your craft must comply with the appropriate standards.

4 Every effort must be made to prevent the connecting flexible cable from falling into the water if it should become disengaged. For this purpose, securing hooks are provided alongside socket-outlets for anchorage at a loop of tie cord.

5 For safety reasons, only one pleasure craft connecting cable supplying one pleasure craft may be connected to any one socket-outlet.

6 The connecting flexible cable must be in one length, without signs of damage , and not contain joints or other means to increase its length.

7 The entry of moisture and salt into the pleasure craft inlet socket may cause a hazard. Examine carefully and clean the plug and socket before connecting the supply.

8 It is dangerous to attempt repairs or alterations. If any difficulty arises, contact the marina management.

BEFORE LEAVING

1 Ensure that the supply is switched off and disconnect all current-using equipment on the craft, before the connecting cable is disconnected and any tie cord loops are unhooked.

2 The connecting flexible cable should be disconnected **firstly** from the marina socket-outlet and **then** from the pleasure craft inlet socket. Any cover that may be provided to protect the inlet from weather should be securely replaced. The connecting flexible cable should be coiled up and stored in a dry location where it will not be damaged.

*   Appropriate figures and colours to be inserted:
nominally 230 V 50 Hz blue – single-phase, and
nominally 400 V 50 Hz red – three-phase.

Alternatively, the notice should be placed in a prominent position or issued to each berth holder. The notice should contain the text of Figure 8.3.

8.11 General notes

8.11.1 Pontoon amenity lighting

It is important that the routes of pontoons and their termination points are clearly delineated.

The lighting may be controlled by either automatic photoelectric cells or time switches, the former being preferred as they sense poor conditions caused by fog, etc. when natural light is waning.

709.512   Luminaires should be of rugged and watertight construction and should preferably be mounted at low level with the light source facing the walkway, not omnidirectional.

8.11.2 Navigation lighting

The local waterway authority should be consulted in order that all necessary and suitably coloured navigation lighting is provided. The light sources should have an extended life expectancy. Photoelectric cell control is preferred to time switches.

8.11.3 Fuelling stations

The relevant local authority should be consulted in order to ensure that the completed installation complies with its requirements. Where applicable, special emergency control facilities should be established onshore. Fuel hoses are required to be non-conducting. Ship/shore bonding cables are not to be used – see *The International Safety Guide for Oil Tankers and Terminals*, 4th Edition.

Electrical equipment in the proximity of fuelling stations should comply with APEA/IP *Guidance for the Design, Construction, Modification, Maintenance and Decommissioning of Filling Stations* (March 2005).

8.11.4 Metering systems

Metering systems are outside the scope of this Guidance Note and must be agreed between the designer and the marina owner to provide all necessary electricity consumption information for accurate billing. The meters may be required to be installed locally in the feeder pillars for local direct reading, or may be part of a site-wide data network system. The metering system must be fit for the installation and type of use. Functional and safety earthing must be adequate.

Check metering for the various main sections of the distribution system may be required in order that the marina operator can use this data in establishing tariffs for the resale of electricity. Such equipment must be installed within the main switchgear and feeder pillars, and must be of adequate rating and quality for the duty required. The increased use of items of electrical equipment exhibiting low power factor characteristics, e.g. dehumidifiers, refrigerators, battery chargers, etc., requires that electricity metering should record suitable data to ensure the marina operator does not suffer a loss of revenue. (This particularly applies when kVAh metering is installed by the electricity distributor.)

8.11.5 Location of equipment

Due consideration should be given to the location of items of equipment so that they are, as far as practicable, not vulnerable to damage either on- or offshore at the marina.

In the case of onshore areas there will be the need for clear vehicular movement including large mobile boat hoists, transit lorries and cars, etc. The location therefore of feeder pillars and lighting columns requires special attention.

For marina areas, the lighting columns and power supply feeder pillars should be so positioned that the risk of contact with luggage trolleys etc. and such items as the bowsprit of craft is, as far as practicable, reduced to a minimum. This is particularly important where lighting and power supply equipment has moulded enclosures which are unable to withstand such mechanical forces and impact and may be damaged.

Site investigations should be carried out at an early stage to determine maximum wave heights which can be experienced. This is of particular importance at exposed coastal sites. Where marinas have breakwater type pontoons, it is likely that under certain conditions waves will pass over the structure.

8.11.6 Routine maintenance and testing

Initial inspection and testing of all electrical systems should be carried out on completion of the installation, in accordance with the requirements of Part 6 of BS 7671 and the recommendations of the IET's Guidance Note 3. A periodic inspection and test of all electrical systems should be carried out annually and the necessary maintenance work implemented. If the site is considered to be exposed, or operational experience shows problems (i.e. misuse), the inspection frequency should be increased to cater for the particular conditions experienced.

GN3

All RCDs should be tested regularly by operating the test button and periodically by a proprietary instrument to ensure they conform with the parameters of their relevant product standards, e.g. BS 4293, BS EN 61008.

All tests should be tabulated for record purposes and the necessary forms required by Part 6 of BS 7671 must be provided by the contractor or persons carrying out the inspection and tests to the person ordering the work.

Medical locations 9

9.1 Introduction and scope

Part 7 of BS 7671:2008 does not include any requirements for medical locations, Section 710 being reserved for this purpose. This chapter is based on the published standard IEC 60364-7-710: 2002 as modified by draft CENELEC standard prHD 60364-7-710 (June 2008) and also takes into account the UK comments on the latter draft submitted to CENELEC.

The particular requirements of prHD 60364-7-710 apply to electrical installations in medical locations so as to ensure safety of patients and medical staff. These requirements, in the main, refer to hospitals, private clinics, medical and dental practices, healthcare centres and dedicated medical rooms in the workplace. The prHD excludes 'medical electrical equipment' as this is covered by the BS EN 60601 series of publications.

The use of medical electrical equipment can be split into three categories; some examples are listed below:

a Life support: infusion pumps, ventilators and dialysis machines etc.
b Diagnostic: X-ray machines, CT scanners, magnetic resonance imagers, blood pressure monitors, electroencephalograph (EEG) and electrocardiograph (ECG) equipment
c Treatment: surgical diathermy and defibrillators etc.

The increased use of this equipment on patients undergoing acute care requires enhanced reliability and safety of the electrical installation in hospitals to ensure the security of supplies and minimise the incidence of electric shock.

For medical locations, Chapter 56 of BS 7671 'Safety services' is particularly relevant. *Chap 56*

9.2 The risks

In medical locations stringent measures are necessary to ensure the safety of patients likely to be subjected to the application of medical electrical equipment.

Shock hazards due to bodily contact with the 50 Hz mains supply are well known and documented. Currents of the order of 10 mA passing through the human body can result in muscular paralysis followed by respiratory paralysis depending on skin resistances, type of contact, environmental conditions and duration. Eventual ventricular fibrillation can occur at currents just exceeding 20 mA. These findings are listed in IEC/TS 60479-1 *Effects of current on human beings and livestock – Part 1: General aspects.*

The natural protection of the human body is considerably reduced when certain clinical procedures are being performed on it. Patients under treatment may have their skin resistance broken or their defensive capacity either reduced by medication or nullified while anaesthetised. These conditions increase the possible consequences of a shock under fault conditions.

In patient environments where intracardiac procedures[1] are undertaken, the electrical safety requirements are even stricter, in order to protect the patient against 'microshock'. Patient leakage currents from applied parts introduced directly to the heart can interfere with cardiac function at current levels which would be considered safe under other circumstances.

Patient leakage current which can flow into an earthed patient is normally greatest when the equipment earth is disconnected. A limit is set to the amount of leakage current which can flow in the patient circuit when the protective earth conductor is disconnected. Patient leakage currents[2] of the order of 10 μA have a probability of 0.2% for causing ventricular fibrillation or pump failure when applied through a small area of the heart. At 50 μA (microshock), the probability of ventricular fibrillation increases to the order of 1%. (Refer to BS EN 60601-1.)

Additional to the consideration of risk from electric shock, some electrical equipment (life-support equipment, surgical equipment) performs such vital functions that loss of supply would pose an unacceptable risk to patients. Medical locations where such equipment is used require secure supplies. This not only has implications for the provision of safety (emergency) power supplies, but also renders some conventional protective measures unsuitable. Hence, for example, when protecting circuits supplying critical medical equipment, restrictions are stipulated on the use of RCDs.

9.3 Definitions

For the purposes of this chapter, the following definitions apply.

9.3.1 Medical location
Location intended for purposes of diagnosis, treatment (including cosmetic treatment), monitoring and care of patients.

Note: The manner in which a room is to be used necessitates some division into separate groups (0, 1 or 2) for differing medical procedures.

9.3.2 Patient
Living being (person or animal) undergoing a medical, surgical or dental procedure. A person under treatment for cosmetic purposes may be considered as a patient.

---

1   A procedure whereby an electrical conductor is placed within the heart of a patient or is likely to come into contact with the heart, such conductor being accessible outside the patient's body. In this context, an electrical conductor includes insulated wires such as cardiac pacing electrodes or intracardiac ECG electrodes, or insulated tubes filled with conducting fluids (catheter).

2   'Patient leakage current': current flowing from a medical electrical equipment applied part via the patient to earth.

9.3.3 Medical electrical equipment (ME equipment)

Electrical equipment having an applied part or transferring energy to or from the patient or detecting such energy transfer to or from the patient and which is

a provided with not more than one connection to a particular supply mains, and
b intended by its manufacturer to be used
- ▶ in the diagnosis, treatment or monitoring of a patient, or
- ▶ for compensation or alleviation of disease, injury or disability.

Note: Medical electrical equipment includes those accessories defined by the manufacturer that are necessary to enable the normal use of the medical electrical equipment.

9.3.4 Applied part

Part of medical electrical equipment that in normal use necessarily comes into physical contact with the patient, in order for medical electrical equipment or a medical electrical system to perform its function.

9.3.5 Group 0

Medical location where no applied parts are intended to be used and where discontinuity (failure) of the supply cannot cause danger to life.

9.3.6 Group 1

Medical location where discontinuity of the electrical supply does not normally represent a threat to the safety of the patient and applied parts are intended to be used as follows:

i externally
ii invasively to any part of the body, except where group 2 applies.

9.3.7 Group 2

Medical location where applied parts are intended to be used in applications where discontinuity (failure) of the supply can cause danger to life, such as:

i intracardiac procedures, or
ii vital treatment and surgical operations.

9.3.8 Medical electrical system (ME system)

Combination, as specified by the manufacturer, of items of equipment, at least one of which is medical electrical equipment to be interconnected by functional connection or by use of a multiple socket-outlet.

Note: The system includes those accessories that are needed for operating the system and are specified by the manufacturer.

9.3.9 Patient environment

Any volume in which intentional or unintentional contact can occur between a patient and parts of the medical electrical equipment or medical electrical system or between a patient and other persons touching parts of the medical electrical equipment or medical electrical system (see Figure 9.1).

▼ **Figure 9.1**
Patient environment

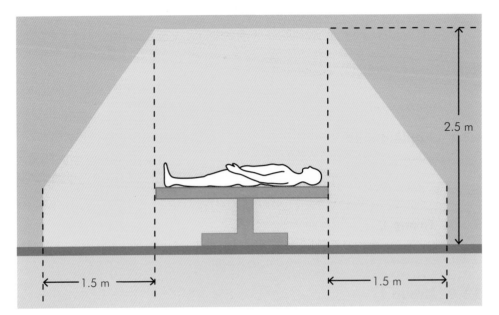

a Elevation

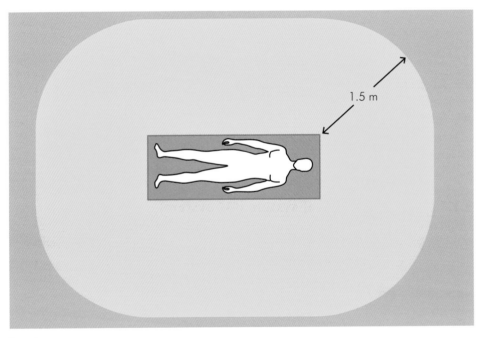

b Plan

Note: This applies when the patient's position is pre-determined; if not, all possible patient positions should be considered.

9.3.10 Medical IT system

IT electrical system fulfilling specific requirements for medical applications.

Note: These supplies are also known as isolated power supply systems.

9.4 Assessment of general characteristics

Allocation of group numbers and classification of safety services of medical locations must be made in agreement with the medical staff and the body responsible for safety. In order to determine the classification of a medical location, it is necessary that the medical staff indicate which medical procedures will take place within the location. Based on the intended use, the appropriate classification for the location can be determined.

Notes:

1 Guidance on the allocation of a group number and classification of safety services for medical locations is shown in Table 9.1.
2 The possibility that certain medical locations could be used for different purposes may require a higher group allocation.

Medical locations[3]		Group			Class	
		0	1	2	≤0.5 s	>0.5 s ≤15 s
1	Massage room	X	X			X
2	Bedrooms		X			
3	Delivery room		X		X[1]	X
4	ECG, EEG, EHG room		X			X
5	Endoscopic room		X[2]			X[2]
6	Examination or treatment room		X			X
7	Urology room		X[2]			X[2]
8	Radiological diagnostic and therapy room, other than mentioned in 21		X			X
9	Hydrotherapy room		X			X
10	Physiotherapy room		X			X
11	Anaesthetic room			X	X[1]	X
12	Operating theatre			X	X[1]	X
13	Operating preparation room		X	X	X[1]	X
14	Operating plaster room		X	X	X[1]	
15	Operating recovery room		X	X	X[1]	X
16	Heart catheterisation room			X	X[1]	X
17	Intensive care room			X	X[1]	X
18	Angiographic examination room			X	X[1]	X
19	Haemodialysis room		X			
20	Magnetic resonance imaging (MRI)		X			X
21	Nuclear medicine		X			X
22	Premature baby room			X	X[1]	X

▼ **Table 9.1**
Examples for allocation of group numbers and classification for safety services of medical locations

Notes:

1 Luminaires and life-support medical electrical equipment which needs power supply within 0.5 s or less.
2 Not being an operating theatre.
3 A definitive list of medical locations showing their assigned groups is impracticable, as the use to which locations (rooms) might be put will differ. This list of examples is provided as a guide only.

9.5   Type of system earthing

The TN-C system is not allowed in medical locations and medical buildings downstream of the main distribution board.

9.6   Power supply

In medical locations the distribution system should be designed and installed to facilitate the automatic changeover from the main distribution network to the electrical safety source feeding essential loads.

9.7   Protection against electric shock

Protective measures providing basic protection (protection against direct contact) only, by obstacles and placing out of reach, are not permitted.

Protective measures for application only when the installation is controlled or under the supervision of skilled or instructed persons, such as non-conducting location and protection by earth-free local equipotential bonding, are not permitted.

Only protection by insulation of live parts, protection by barriers or enclosures or by the use of Class II equipment are permitted.

9.8   Requirements for fault protection (protection against indirect contact)

9.8.1   Automatic disconnection in case of a fault

Care shall be taken to ensure that simultaneous use of many items of equipment connected to the same circuit cannot cause unwanted tripping of the residual current protective device (RCD) where fitted.

In medical locations of group 1 and group 2, where RCDs are required, only type A or type B shall be selected depending on the possible fault current arising. Type AC RCDs shall not be used.

Where medical IT systems are used, additional protection by means of an RCD is not permitted.

For IT, TN and TT systems in medical locations of group 1 and group 2, the voltage between simultaneously accessible exposed-conductive- and/or extraneous-conductive-parts shall not exceed 25 V a.c. or 60 V d.c.

▼ **Table 9.2**
Maximum disconnection times for TN and TT systems in group 1 and 2 locations

Note: For TN and TT systems, Table 9.2 gives guidance on the recommended disconnection times.

System	$25\text{ V} < U_0 \le 50\text{ V}$ (seconds)		$50\text{ V} < U_0 \le 120\text{ V}$ (seconds)		$120\text{ V} < U_0 \le 230\text{ V}$ (seconds)		$230\text{ V} < U_0 \le 400\text{ V}$ (seconds)		$U_0 > 400\text{ V}$ (seconds)	
	a.c.	d.c.	a.c.	d.c.	a.c.	d.c.	a.c.	d.c.	a.c.	d.c.
TN	5	5	0.3	2	0.3	0.5	0.05	0.06	0.02	0.02
TT	5	5	0.15	0.2	0.05	0.1	0.02	0.06	0.02	0.02

9.8.2   TN systems

In final circuits of group 1 locations rated up to 63 A, RCDs with a rated residual operating current not exceeding 30 mA and meeting the characteristics of Regulation 415.1.1 shall be used. 415.1.1

In medical locations of group 2 (except for the medical IT system), protection by automatic disconnection of supply by means of RCDs with a rated residual operating current not exceeding 30 mA shall only be used on the following circuits:

▶ circuits for the supply of movements of operating tables
▶ circuits for X-ray units (**Note:** The requirement is mainly applicable to mobile X-ray units brought into group 2 locations)
▶ circuits for large equipment with a rated power greater than 5 kVA
▶ circuits for non-critical electrical equipment (non life-support).

In circuits designed for rated current above 63 A, RCDs with a rated residual operating current not exceeding 300 mA shall be used.

Note: It is recommended that TN-S systems are monitored to ensure the insulation level of all live conductors.

9.8.3   TT systems

In medical locations of group 1 and group 2, the requirements of TN systems (see 9.8.2) apply and RCDs shall be used as disconnection devices.

9.9   Medical IT system

In medical locations of group 2, the medical IT system shall be used for circuits supplying medical electrical equipment and systems intended for life support, surgical applications and other electrical equipment located in the 'patient environment', excluding equipment listed in section 9.8.2 (see Figure 9.2).

For each group of rooms serving the same function, at least one separate medical IT system is necessary. The medical IT system shall be equipped with an insulation monitoring device (IMD) in accordance with BS EN 61557-8, with the following specific requirements:

▶ a.c. internal impedance shall be $\geq$100 kΩ
▶ internal resistance shall be $\geq$250 kΩ
▶ test voltage shall not be greater than 25 V d.c.
▶ injected current, even under fault conditions, shall be $\leq$1 mA peak
▶ indication shall take place at the latest when the insulation resistance has decreased to 50 kΩ. If the response value is adjustable, the lowest possible set-point value shall be 50 kΩ. A test device shall be provided
▶ response and alarm-off time shall be $\leq$5 s.

Notes:
1   An indication is recommended if the earth continuity of the IMD is lost
2   The necessary additional requirements for IMDs given above are at this time not covered in the equipment standard BS EN 61557-8. They will be removed from this publication following their inclusion in the relevant equipment standard.

▼ **Figure 9.2** Typical medical IT system with insulation monitoring

a Theatre suite

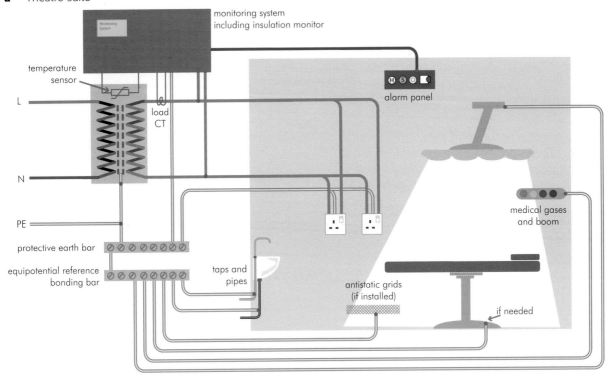

b Network illustrating two separate distribution circuits

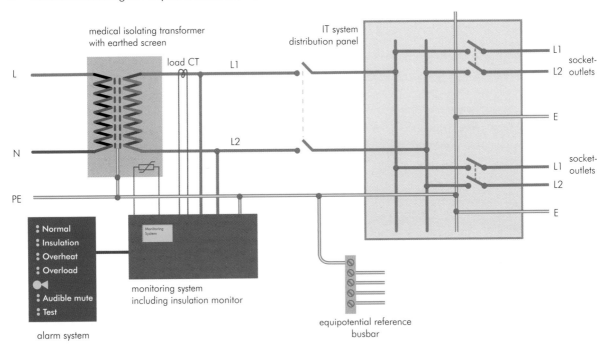

For each medical IT system, an acoustic and visual alarm system incorporating the following components shall be arranged at a suitable place so that it can be permanently monitored (audible and visual signals) by the medical staff:

▶ a green signal lamp to indicate normal operation
▶ a yellow signal lamp which lights when the minimum value set for the insulation resistance is reached – it shall not be possible for this light to be cancelled or disconnected
▶ an audible alarm which sounds when the minimum value set for the insulation resistance is reached – this audible alarm may be silenced
▶ the yellow signal shall go out on removal of the fault and when the normal condition is restored.

Monitoring of overload and high temperature for the medical IT transformer is required.

9.10 Transformers for medical IT systems

Transformers shall be in accordance with BS EN 61558-2-15, with the following additional requirements.

The leakage current of the output winding to earth and the leakage current of the enclosure, when measured under no-load conditions and with the transformer supplied at rated voltage and rated frequency, shall not exceed 0.5 mA.

Single-phase transformers shall be used to form the medical IT systems for mobile and fixed equipment and the rated output shall not be less than 0.5 kVA and shall not exceed 10 kVA.

If the supply of three-phase loads via an IT system is also required, a separate three-phase transformer shall be provided for this purpose with an output line-to-line voltage not exceeding 250 V.

9.11 Protection by SELV or PELV

When using SELV and/or PELV circuits in medical locations of group 1 and group 2, the nominal voltage applied to current-using equipment shall not exceed 25 V rms a.c. or 60 V ripple-free d.c.

Protection by functional extra-low voltage (FELV) is not permitted.

Protection by basic insulation of live parts and by barriers or enclosures is essential.

In medical locations of group 2, where PELV is used, exposed-conductive-parts of equipment (e.g. operating theatre luminaires) shall be connected to the protective (equipotential) bonding conductor.

9.12 Supplementary equipotential bonding

9.12.1 Group 1 and 2 medical locations

In each medical location of group 1 and group 2, supplementary bonding conductors shall be installed and connected to the equipotential bonding busbar for the purpose of equalising potential differences between the following parts located in the 'patient environment':

▶ protective conductors
▶ extraneous-conductive-parts
▶ screening against electrical interference fields, if installed
▶ connection to conductive floor grids, if installed
▶ metal screen of isolating transformers, via the shortest route to the protective conductor.

A sufficient number of supplementary equipotential bonding connection points for the connection of medical electrical equipment shall be available in group 2 locations and recommended in group 1 locations.

Note: Fixed conductive non-electrical patient supports, such as operating theatre tables, physiotherapy couches and dental chairs, should be connected to the protective (equipotential) bonding conductor unless they are intended to be isolated from Earth.

9.12.2 Resistance of protective conductors in group 1 locations

In medical locations of group 1, the measured resistance of protective conductors, including the resistance of the connections, between the terminals for the protective conductor of socket-outlets and of fixed equipment or any extraneous-conductive-parts and the equipotential bonding busbar shall not exceed 0.7 ohm.

Note: The resistance value of 0.7 ohm is associated with the disconnection times of Table 9.2.

9.12.3 Resistance of protective conductors in group 2 locations

In medical locations of group 2, the measured resistance of protective conductors, including the resistance of the connections, between the terminals for the protective conductor of socket-outlets and of fixed equipment or any extraneous-conductive-parts and the equipotential bonding busbar shall not exceed 0.2 ohm.

9.12.4 Equipotential bonding busbar

The equipotential bonding busbar shall be located in or near the medical location. In each distribution board or in its proximity, an additional equipotential bonding bar shall be provided to which the supplementary bonding conductors and protective earth conductor shall be connected. Connections shall be so arranged that they are accessible, labelled, clearly visible and can readily be disconnected individually.

9.13 Distribution boards

Distribution boards shall be in accordance with BS EN 60439 series.

Distribution boards for group 2 medical locations should be installed in close proximity to that location and clearly identified.

Note: Distribution boards should preferably be installed outside medically used locations and be safely guarded against unauthorised access.

9.14 Explosion risk

Electrical devices (e.g. socket-outlets and switches) shall be installed at a distance of at least 0.2 m horizontally (centre to centre) but not directly below any medical gas-outlets for oxidising or flammable gases, so as to minimise the risk of ignition of flammable gases.

Note: Requirements for medical electrical equipment for use in conjunction with flammable gases and vapours are contained in BS EN 60601-1.

9.15 Wiring systems in group 2 medical locations

Any wiring system within group 2 medical locations shall be exclusive to the use of equipment and fittings in that location.

Overload current protection is not permitted in the output line (secondary) downstream of the medical IT transformer. Fuses may be used for short-circuit protection only.

9.16 Lighting circuits

In medical locations of group 1 and group 2, at least two different sources of supply shall be provided for some of the luminaires by two separate circuits. One of the two circuits shall be connected to the safety service.

In escape routes, alternate luminaires shall be connected to the safety service.

9.17 Socket-outlet circuits in the medical IT system for group 2 medical locations

Socket-outlets intended to supply medical electrical equipment shall be unswitched and be fitted with a supply indicator.

At each patient's place of treatment, e.g. bedheads, socket-outlets shall be configured as follows:

▶ each socket-outlet supplied by an individually protected circuit; or
▶ several socket-outlets supplied by a minimum of two separate circuits. See Figure 9.2b (**Note:** it is recommended that each individual circuit is monitored for overload and earth fault).

Where circuits are supplied from other systems (TN-S or TT) in the same medical location, socket-outlets connected to the medical IT system shall be either:

▶ of such construction that prevents their use in other systems, or
▶ clearly and permanently marked.

9.18 Safety services

In medical locations, a power supply for safety services is required which, in accordance with the standard, will energize the installations needed for continuous operation in case of failure of the general power system for a defined period with a preset changeover time.

The safety power supply system shall automatically take over, if the voltage of one or more incoming live conductors to the main distribution board has dropped for more than an agreed time and by more than an agreed percentage in regard to the nominal voltage. Classification of safety services is given in Table 9.3.

Note: The responsible management of the medical location (including medical staff) should be involved in the decision where safety services are needed.

▼ **Table 9.3**
Classification of safety services necessary for medical locations

Classification	Changeover time (s)	Description
No break	0	Automatic supply available at no-break
Very short break	0.15	Automatic supply available within 0.15 s
Short break	0.5	Automatic supply available within 0.5 s
Medium break	15	Automatic supply available within 15 s
Long break	>15	Automatic supply available in more than 15 s

Notes:
1 Generally it is unnecessary to provide a no-break power supply for medical electrical equipment. However, certain microprocessor controlled equipment may require such a supply.
2 Safety services provided for locations having differing classifications should meet that classification which gives the highest security of supply. Refer to Table 9.1 for guidance on the association of classification of safety services with medical locations.
3 The notation 'within' implies '$\leq$'.

9.18.1 General requirements for safety power supply sources of groups 1 and 2
An additional main incoming power supply from the general power supply is not regarded as a source of the safety power supply.

Notes:
1 If unit-type power stations with jacking-piston combustion engines are used as safety power sources, see ISO 8528. For calculations of supplied power, only prime power (PRP) specifications shall be used (see ISO 8528-1).
2 The availability (readiness for service) of safety power supply sources is to be monitored and indicated at a suitable location.

9.18.2 Failure of the general power supply source
In case of a failure of the general power supply source, the power supply for safety services shall be energized to feed the equipment stated in sections 9.18.5, 9.18.6 and 9.18.7 with electrical energy for a defined period of time and within a pre-determined changeover period.

9.18.3 Interconnecting cables of safety power supply sources

For interconnecting cables between the individual components and the subassemblies of safety power supply sources, see section 9.15.

Note: The circuit which connects the power supply source for the safety services to the main distribution board should be considered a safety circuit.   560.8.3

9.18.4 Socket-outlets

Where socket-outlets are supplied from the safety power supply source they shall be readily identified.

9.18.5 Power supply sources with a changeover period ≤0.5 s

In the event of a voltage failure of one or more line conductors at the distribution board, a special safety power supply source shall be used capable of providing power for a minimum of 3 h for:

▶ luminaires of operating theatre tables
▶ other essential luminaires, e.g. endoscopes
▶ critical life-supporting medical electrical equipment.

Notes:
1 The duration of 3 h may be reduced to 1 h if a power source according to section 9.18.6 is installed and the essential luminaires for operation can be supplied from the source.
2 Included under other essential luminaires may be light sources for endoscopic surgical field-luminaires.

9.18.6 Power supply sources with a changeover period ≤15 s

Equipment according to sections 9.18.8 and 9.18.9 shall be connected within 15 s to a safety power supply source capable of maintaining it for a minimum period of 24 h, when the voltage of one or more live conductors at the main distribution board for the safety services has decreased by more than 10% of the nominal value of supply voltage and for a duration greater than 3 s.

Note: The duration of 24 h can be reduced to a minimum of 3 h if the medical requirements and the use of the location, including any treatment, can be concluded and if the building can be evacuated in a time which is well within 24 h.

9.18.7 Power supply sources with a changeover period >15 s

Equipment other than that covered by sections 9.18.5 and 9.18.6, which is required for the maintenance of hospital services, may be connected either automatically or manually to a safety power supply source capable of maintaining it for a minimum period of 24 h.

Note: This equipment may include, for example: sterilisation equipment, technical building installations, in particular air conditioning, heating and ventilation systems, building services and waste disposal systems, cooling equipment, storage battery chargers.

9.18.8 Safety lighting

In the event of mains power failure, the necessary minimum illuminance shall be provided from the safety services source for the following locations:

▶ escape routes (arranged in alternate circuits)
▶ lighting of exit signs
▶ locations for switchgear and controlgear for emergency generation sets and for main distribution boards of the normal power supply and for power supply for safety services
▶ rooms in which essential services are intended. In each room at least one luminaire shall be supplied from the power source for safety services
▶ locations of central fire alarm and monitoring systems
▶ rooms of group 1 medical locations – in each room at least one luminaire shall be supplied from the power supply source for safety services
▶ rooms of group 2 medical locations – a minimum of 50% of the lighting shall be supplied from the power source for safety services.

Note: The values for minimum illuminance can be given by national and/or local regulations.

9.18.9 Other services

Services other than lighting which require a safety service supply may include, for example, the following:

▶ selected lifts for firefighters
▶ ventilation systems for smoke extraction
▶ paging systems
▶ medical electrical equipment used in group 2 medical locations which serves for surgical or other measures of vital importance; such equipment will be defined by responsible staff
▶ electrical equipment of medical gas supply including compressed air, vacuum supply and narcosis (anaesthetics) exhaustion as well as their monitoring devices
▶ fire detection, fire alarms and fire extinguishing systems.

9.19 Inspection and testing

Chap 63   The dates and results of each initial verification and periodic inspection and testing should be recorded in the form specified in Chapter 63 of BS 7671:2008.

9.19.1 Initial verification

Chap 61   The tests specified below under items **a** to **g** in addition to the requirements of Chapter 61 of BS 7671:2008 shall be carried out, both prior to commissioning and after alterations or repairs and before re-commissioning.

a Functional test of insulation monitoring devices of medical IT systems and acoustical/visual alarm systems.
b Measurements to verify that the supplementary equipotential bonding is in accordance with sections 9.12.2 and 9.12.3.
c Verification of the integrity of the facilities required for equipotential bonding.
d Verification of the integrity of the requirements of section 9.18 for safety services.

e Measurements of leakage current of the output circuit and of the enclosure of medical IT transformers in no-load condition.

f Computational verification of the compliance of the selectivity of the safety power supply with regard to planning documents and calculation.

g Verification of the applied protective measures for compliance with the requirements for group 1 and group 2.

9.19.2 Periodic inspection and testing

The tests specified below under items **a** to **i**, in addition to the requirements of Chapter 62 of BS 7671:2008, shall be carried out at the stated intervals.

Chap 62

a Functional testing of changeover devices: 12 months

b Functional testing of insulation monitoring devices: 12 months

c Measurement verifying the supplementary equipotential bonding: 36 months

d Verifying integrity of facilities required for equipotential bonding: 36 months

e Monthly functional testing of:
  ▸ safety services with batteries: 15 min
  ▸ safety services with combustion engines: 60 min

f Annual functional testing of:
  ▸ safety services with combustion engines, until rated running temperature is achieved: 'endurance run'
  ▸ safety services with batteries: capacity test

  In all cases of **e** and **f** at least 50% to 100% of the rated power shall be taken over.

g Checking of the tripping of RCDs at $I_{\Delta n}$: 12 months

h Visual inspection, functional tests and measurements of the electrical installation, especially to verify the protection against electric shock, including the settings of adjustable protective devices: 36 months

i Functional test of the lighting of exit signs, escape routes, locations for switchgear and controlgear: 12 months

9.20 Recommendations from the Department of Health

The Department of Health provides medical building and engineering guidance through its Health Technical Memoranda (HTM) series of documents. These have recently been revised and five documents are specifically related to electrical engineering issues.

The electrical series of documents are supported by the general policies and procedures embraced in HTM 00. These documents supersede HTMs 2007, 2011, 2014, 2020 and 2021. The current titles are:

HTM 00	Policies and Procedures (2006)
HTM 06-01 Part A	Electrical Services supply and distribution; Design considerations (2007)
HTM 06-01 Part B	Electrical Services supply and distribution; Operational management (2007)
HTM 06-02	Electrical Safety Guidance for low voltage systems (2006)
HTM 06-02	Electrical Safety Handbook
HTM 06-03	Electrical Safety Guidance for high voltage systems (2006)

These documents are supported by a range of standard procedure forms covering the following issues:

▶ Isolation and Earthing Diagram
▶ Safety Programme
▶ Permit-to-work
▶ Limitation-of-access
▶ Sanction-for-test (HV)
▶ Certificate of authorisation for live working (LV)
▶ Permission for disconnection/interruption of electrical services (LV)
▶ Logbook

All of the above documents are available (at cost) from The Stationery Office (TSO) www.tso.co.uk.

HTM 06-01 Part A provides guidance for all works on the fixed wiring and integral electrical equipment used for electrical services within healthcare premises. The document should be used for all forms of electrical design work ranging from a new greenfield site to modifying an existing final circuit.

HTM 06-01 Part B addresses the operational management and maintenance of the electrical services supply distribution within a healthcare facility.

HTM 06-02, together with its associated *Safety Guidance Handbook*, gives operational guidance on electrical safety requirements for low voltage systems in healthcare premises.

HTM 06-03 gives operational guidance on electrical safety requirements for high voltage systems in healthcare premises.

Solar photovoltaic (PV) power supply systems

10

10.1 Introduction

This is a new section that has been introduced into the 17th Edition. There are an increasing number of solar power supplies that are being installed in all kinds of premises, with their own particular risks for installers, users and maintenance personnel.

Section 712

The section includes requirements for accessibility, external influences, routing of protective conductors, selection and erection of cables to avoid the risk of lightning strike, short-circuit and earth faults, overcurrent protection, compliance with standards, protection against electromagnetic interference, and devices for isolation.

10.2 Scope

The particular requirements of Section 712 apply to electrical installations of PV power supply systems including systems with a.c. modules.

712.1

Note: Requirements for PV power supply systems which are intended for stand-alone operation are under consideration by the IEC.

10.3 The Electricity Safety, Quality and Continuity Regulations 2002

The solar photovoltaic (PV) power supply systems described in this chapter are required to meet the requirements of the Electricity Safety, Quality and Continuity Regulations 2002 as they are embedded generators.

Where the output does not exceed 16 A per phase they are by definition small-scale embedded generators (SSEG) and are exempted from certain of the requirements of Regulation 22 provided that:

i   the equipment should be type tested and approved by a recognised body,
ii   the consumer's installation must comply with the requirements of BS 7671,
iii the equipment must disconnect itself from the distributor's network in the event of a network fault, and
iv   the distributor must be advised of the installation before or at the time of commissioning.

The requirements of the ESQC Regulations for small-scale embedded generators are discussed more fully in Chapter 15.

Installations will need to meet the requirements of the Energy Networks Association (formerly the Electricity Association) Engineering Recommendation G83, *Recommendations for the connection of small-scale embedded generators (up to 16 A per phase) in parallel with the public low voltage distribution network.*

10.4 Normative references

The following IEC standards are referred to in publication G83:

▶ IEC 60904-3: *Photovoltaic devices – Part 3: Measurement principles for terrestrial photovoltaic (PV) solar devices with reference spectral irradiance data*
▶ IEC 612215: *Crystalline silicon terrestrial photovoltaic (PV) modules – Design qualification and type approval.*

Standards for equipment are being developed by IEC Committee TC 82.

10.5 Definitions

(See also Figures 10.1 and 10.2.)

Part 2

PV cell: basic PV device which can generate electricity when exposed to light such as solar radiation.

PV module: smallest completely environmentally protected assembly of interconnected PV cells.

PV string: circuit in which PV modules are connected in series, in order for a PV array to generate the required output voltage.

PV array: mechanically and electrically integrated assembly of PV modules, and other necessary components, to form a d.c. power supply unit.

PV array junction box: enclosure where all PV strings of any PV array are electrically connected and where devices can be located.

PV generator: assembly of PV arrays.

PV generator junction box: enclosure where all PV arrays are electrically connected and where devices can be located.

PV string cable: cable connecting PV modules to form a PV string.

PV array cable: output cable of a PV array.

PV d.c. main cable: cable connecting the PV generator junction box to the d.c. terminals of the PV invertor.

PV invertor: device which converts d.c. voltage and d.c. current into a.c. voltage and a.c. current.

PV supply cable: cable connecting the a.c. terminals of the PV invertor to a distribution circuit of the electrical installation.

PV a.c. module: integrated module/invertor assembly where the electrical interface terminals are a.c. only. No access is provided to the d.c. side.

PV installation: erected equipment of a PV power supply system.

Standard test conditions (STC): test conditions specified in BS EN 60904-3 for PV cells and modules.

Open-circuit voltage under standard test conditions U_{OC} STC: voltage under standard test conditions across an unloaded (open) PV module, PV string, PV array, PV generator, or on the d.c. side of the PV invertor.

Short-circuit current under standard test conditions I_{SC} STC: short-circuit current of a PV module, PV string, PV array or PV generator under standard test conditions.

DC side (not a BS 7671 definition): part of a PV installation from a PV cell to the d.c. terminals of the PV invertor.

AC side (not a BS 7671 definition): part of a PV installation from the a.c. terminals of the PV invertor to the point of connection of the PV supply cable to the electrical installation.

Simple separation: separation between circuits or between a circuit and Earth by means of basic insulation.

10.6 Protection for safety

10.6.1 Protection against electric shock

PV equipment on the d.c. side is to be considered energized, even when the system is disconnected from the a.c. side. 712.410.3

The selection and erection of equipment shall facilitate safe maintenance and shall not adversely affect provisions made by the manufacturer of the PV equipment to enable maintenance or service work to be carried out safely. 712.513.1

Protection by extra-low voltage: SELV and PELV 712.414.1.1
For SELV and PELV systems, U_{OC} STC replaces U_0 and shall not exceed 120 V d.c.

10.6.2 Fault protection

(a) Protection by automatic disconnection of supply 712.411
Note: Protection by automatic disconnection of supply on the d.c. side requires special measures which are under consideration.

On the a.c. side, the PV supply cable shall be connected to the supply side of the protective device for automatic disconnection of circuits supplying current-using equipment.

Where an electrical installation includes a PV power supply system without at least simple separation between the a.c. side and the d.c. side, an RCD installed to provide fault protection by automatic disconnection of supply shall be type B to IEC 60755 amendment 2.

Where the PV power supply invertor by construction is not able to feed d.c. fault currents into the electrical installation, an RCD of type B to IEC 60755 amendment 2 is not required.

(b) Other protective measures
Protection by use of Class II or equivalent insulation should preferably be adopted on the d.c. side. 712.412

Protection by non-conducting location (Regulation 418.1) is not permitted on the d.c. side. 712.410.3.6

Protection by earth-free local equipotential bonding (Regulation 418.2) is not permitted on the d.c. side.

10.6.3 Protection against overload on the d.c. side

Note: For PV cables on the d.c. side not complying with the paragraphs below, the requirements of Chapter 43 of BS 7671 apply for overload protection.

712.433.1 Overload protection may be omitted to PV string and PV array cables when the continuous current-carrying capacity of the cable is equal to or greater than 1.25 times I_{SC} STC at any location.

712.433.2 Overload protection may be omitted to PV main cables when the continuous current-carrying capacity of the PV main cable is equal to or greater than 1.25 times I_{SC} STC of the PV generator.

Note: These requirements are only relevant for protection of the cables. See also the manufacturer's instructions for protection of the PV modules.

10.6.4 Protection against fault current

712.434.1 The PV supply cable shall be protected against fault current by an overcurrent protective device installed at the connection to the a.c. mains.

10.6.5 Protection against electromagnetic interference (EMI)

712.444.4.4 To minimise voltages induced by lightning, the area of all wiring loops shall be kept as small as possible.

10.7 Isolation and switching

712.537.2.1.1 To allow maintenance of the invertor, means of isolating the PV invertor from the d.c. side and the a.c. side shall be provided.

Note: Further requirements with regard to the isolation of a PV installation operating in parallel with the public supply system are given in Regulation 551.7.6 of BS 7671.

10.8 Selection and erection of electrical equipment

10.8.1 Compliance with standards

712.511.1 PV modules shall comply with the requirements of the relevant equipment standard, e.g. BS EN 612215 for crystalline PV modules. PV modules of Class II construction or with equivalent insulation are recommended if U_{OC} STC of the PV strings exceeds 120 V d.c.

The PV array junction box, PV generator junction box and switchgear assemblies shall be in compliance with BS EN 60439-1.

10.8.2 Operational conditions and external influences

712.512 Electrical equipment on the d.c. side shall be suitable for direct voltage and direct current.

PV modules may be connected in series up to the maximum allowed operating voltage of the PV modules or the PV invertor, whichever is lower. Specifications for this equipment shall be obtained from the equipment manufacturer.

If blocking diodes are used, their reverse voltage shall be rated for 2 x U_{OC} STC of the PV string. The blocking diodes shall be connected in series with the PV strings (see Figures 10.1 and 10.2).

When installing PV modules the installer must follow the manufacturer's instructions for mounting so that adequate heat dissipation is provided under the conditions of maximum solar radiation to be expected. Such instructions are required by the equipment standard.

712.512.2.1

10.8.3 Wiring systems

Selection and erection in relation to external influences

712.522

PV string cables, PV array cables and PV d.c. main cables shall be selected and erected so as to minimise the risk of earth faults and short-circuits.

Note: This may be achieved for example by reinforcing the protection of the wiring against external influences by the use of single-core sheathed cables.

Wiring systems shall withstand the expected external influences such as wind, ice formation, temperature and solar radiation.

10.8.4 Devices for isolation and switching

In the selection and erection of devices for isolation and switching to be installed between the PV installation and the public supply, the public supply shall be considered the source and the PV installation shall be considered the load. A notice warning of two sources of supply should be provided at each point of isolation.

712.537.2.2

514.15.1

A switch disconnector shall be provided on the d.c. side of the PV invertor.

All junction boxes (PV generator and PV array boxes) shall carry a warning label indicating that active parts inside the boxes may still be live after isolation from the PV invertor.

The Department of Energy and Climate Change (DECC) (formerly the DTI) guidance on the ESQC Regulations advise that the means of disconnection should preferably be by mechanical separation of contacts. However, a suitably rated solid-state switching device is permitted by DECC provided it is equipped with fail-safe monitoring to ensure that the line to neutral voltage on the mains side of the device reduces to less than 50 volts within 0.5 s of the device failing to operate when required to do so. The means of isolation of the generating plant (for the purposes of working either on the consumer's system or the distributor's network as required) should be by an accessible all-lines and neutral manually operated electromechanical isolating switch in all circumstances.

10.9 Earthing arrangements and protective conductors

Where protective bonding conductors are installed, they shall be parallel to and in as close contact as possible with d.c. cables and a.c. cables and accessories.

712.54

10.9.1 Types of system earthing

Earthing of one of the live conductors of the d.c. side is permitted, if there is at least simple separation between the a.c. side and the d.c. side.

712.312.2

Note: Any connections with Earth on the d.c. side should be electrically connected so as to avoid corrosion.

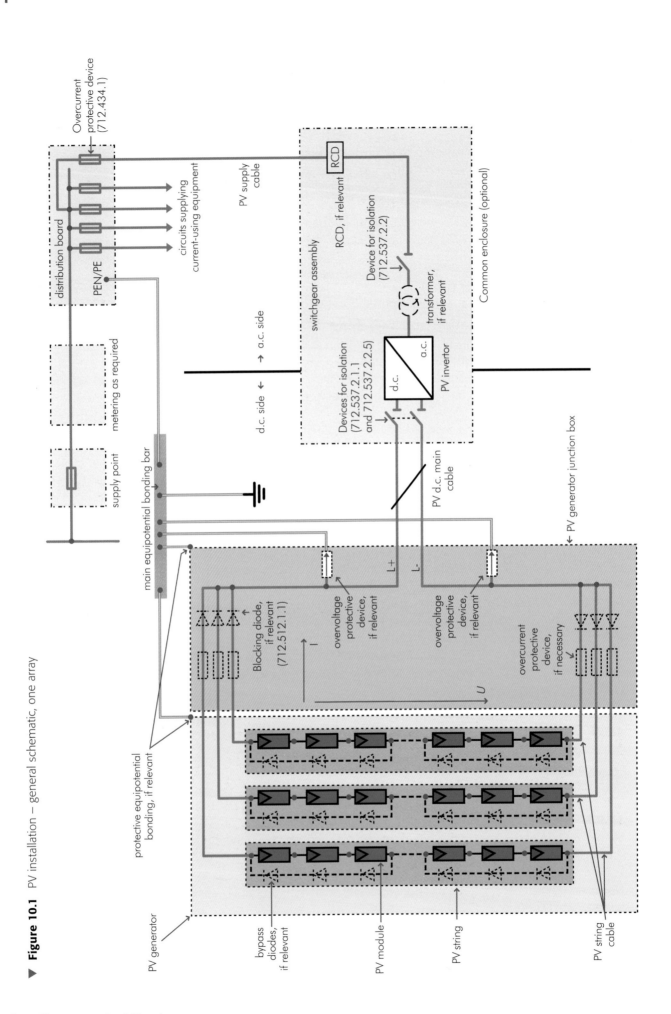

▶ **Figure 10.1** PV installation – general schematic, one array

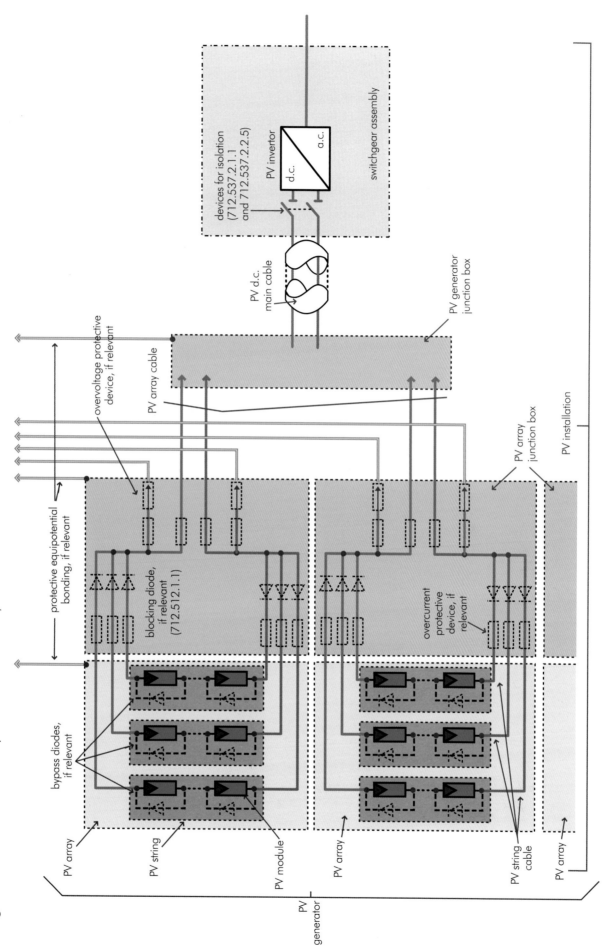

▶ **Figure 10.2** PV installation – example with two or more arrays

Exhibitions, shows and stands

11

11.1 Introduction

The 16th Edition did not include requirements for exhibitions, shows and stands, but guidance was given in the previous edition of this Guidance Note. However, new Section 711 concerning the electrical installations of exhibitions, etc. has been introduced into the 17th Edition.

Section 711

Section 711 specifies particular requirements for this type of electrical installation relating to assessment of general characteristics, protection against electric shock, protection against thermal effects, selection and erection of equipment and inspection and testing.

11.2 Scope

This chapter is concerned with temporary electrical installations in exhibitions, shows and stands including mobile and portable displays and equipment. Such installations may be installed indoors or outdoors within permanent or temporary structures. It does not apply to the fixed electrical installation of the building, if any, in which the exhibition, show or stand takes place.

Part 2

Section 711 of BS 7671:2008 does not apply to electrical systems as defined in BS 7909 used in structures, sets, mobile units etc. as used for public or private events, touring shows, theatrical, radio, TV or film productions and similar activities of the entertainment industry.

11.3 The risks

The particular risks associated with exhibitions, shows and stands are those of electric shock and fire. These arise from:

1 the temporary nature of the installation
2 lack of permanent structures
3 severe mechanical stresses
4 access to the general public.

Because of these increased risks, additional measures are recommended as described in the following sections.

11.4 Protection against electric shock

711.410.3.4 A cable intended to supply temporary structures shall be protected at its origin by an RCD whose rated residual operating current does not exceed 300 mA. This device shall provide a delay by using a device in accordance with BS EN 60947-2, or be of the S-type in accordance with BS EN 61008-1 or BS EN 61009-1 for discrimination with RCDs protecting final circuits. See Figure 11.1.

Note: The requirement for additional protection relates to the increased risk of damage to cables in temporary locations.

711.410.3.5 Protective measures against electric shock by means of obstacles and by placing out of reach (Section 417) are not permitted.

711.410.3.6 The protective measures of non-conducting location (Regulation 418.1) and earth-free local equipotential bonding (Regulation 418.2) are not permitted.

11.4.1 Protection by automatic disconnection of supply

711.411.4 Because of the practical difficulties of bonding all accessible extraneous-conductive-parts, a TN-C-S (PME) system is not appropriate for temporary and/or outdoor installations. A TN-S system would be acceptable if such a supply were available from the distributor. It is most likely and preferable for TT systems to be adopted. The ESQC Regulations prohibit the use of TN-C-S systems for the supply to a caravan or similar construction.

11.4.2 Additional protection

711.411.3.3 Each socket-outlet circuit not exceeding 32 A and all final circuits other than for emergency lighting shall be protected by an RCD having the characteristics specified in Regulation 415.1.1. However, consideration has to be given to the hazards of loss of lighting in such a public place, particularly when crowded. Lighting of such areas should always be on at least two separate circuits with separate RCDs, and should preferably be out of reach of the general public (see Figure 11.1).

▼ **Figure 11.1**
Exhibition/show distribution with standby generator

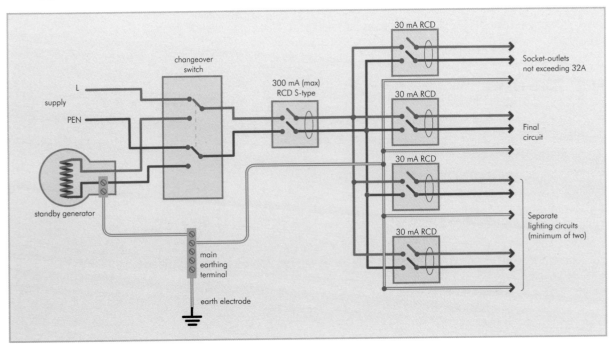

Note: Regulation 21 of the ESQC Regulations has requirements for switched alternative sources of energy; see section 11.9.

11.4.3 Bonding of vehicles, wagons, caravans and stands

Structural metallic parts which are accessible from within the stand, vehicle, wagon, caravan or container shall be connected through the main protective bonding conductors to the main earthing terminal within the unit.

711.411.3.1.2

11.4.4 Protection by SELV or PELV

Where SELV or PELV installations are used, whatever the nominal voltage, basic protection is required to be provided by basic insulation or by barriers or enclosures, providing protection of at least IP4X or IPXXD.

711.414.4.5

11.5 Protection against thermal effects

There is often an increased risk of fire and burns in temporary locations. For this reason it is important to comply with all the relevant requirements of Chapter 42.

Chap 42

11.5.1 Protection against fire

Installation designers must bear in mind that stored materials may present a particular hazard in such a location, particularly fodder, cardboard boxes etc.

Motors which are automatically or remotely controlled and are not continuously supervised should be fitted with manual reset devices for protection against excess temperature, accessible only to skilled persons.

422.3.7
552.1.2

11.5.2 Lighting

Lighting equipment such as incandescent lamps, spotlights and small projectors, and other equipment or appliances with high temperature surfaces should, in addition to being suitably guarded, be arranged well away from combustible materials such as to prevent contact. Equipment should be installed and located in accordance with relevant standards and manufacturers' instructions.

711.422.4.2
422.3.1
Section 559

Luminaires that are mounted below a height of 2.5 m from floor or ground level or mounted in a position that is accessible to accidental contact, should be firmly and adequately fixed, and sited or guarded so as to prevent the risk of injury to persons or ignition of materials. Where this includes outdoor lighting installations the requirements of Regulation group 559.10 would apply.

711.559.5

Showcases and signs should be constructed from materials having adequate heat resistance, mechanical strength, electrical insulation and ventilation, taking account of the combustibility of exhibits in relation to the heat generated. The manufacturer's instructions must be complied with.

Where there is a concentration of electrical equipment, including luminaires, that might generate considerable heat, adequate ventilation must be provided.

Similar guidance for protection against fire for this type of location can be found in section 8.3 of Guidance Note 4: *Protection Against Fire*.

GN4

11.6 Isolation

Every separate temporary structure, such as vehicles, stands or units, intended to be occupied by one specific user, and each distribution circuit supplying outdoor installations, should be provided with its own readily accessible and properly identifiable means of isolation. Switches, circuit-breakers and residual current devices

711.537.2.3
537.2

etc. considered suitable for isolation by the relevant standard or the manufacturer may be used.

11.7 Measures of protection against overcurrent

430.3 All circuits should be protected against overcurrent by a suitable protective device located at the origin of the circuit.

11.8 Selection and erection of equipment

711.51

Part 2 Control and protective switchgear should be placed in closed cabinets which can only be opened by a key or tool, except those parts which are designed and intended to be operated by ordinary persons.

512.2 Equipment, particularly switchgear and fusegear, must be mounted away from locations that may not be weatherproof. Tent poles etc. where used for mounting switchgear are often the weak point in the weather tightness of temporary structures.

The means of isolation for each stand or unit should not be locked away and should be readily accessible and obvious to the stand user (see section 11.6).

11.8.1 Wiring systems

Section 522
Section 523 Particular care must be paid to the selection and installation of cables to ensure that the mechanical protection, insulation, heat resistance and current-carrying capacity are sufficient for the conditions likely to be encountered.

711.52 Mechanical protection or armoured cables should be used wherever there is a risk of damage, and flexible cords should not be laid in areas accessible to the public unless they are protected against mechanical damage.

Underground cables are susceptible to damage by structure support pins which may be up to 1 m in length. Exhibitors must be advised of the presence of cables and if necessary the cable route marked. The general rules for buried cables must be followed; see also section 13.3 in Chapter 13.

711.521 Where a building is used for exhibitions etc. and does not include a fire alarm system, either of the following cable systems should be used:

▶ flame retardant cables to BS EN 60332-1-2 or to a relevant part of the BS EN 50266 series, and low smoke to BS EN 61034-2
▶ single-core or multicore cables enclosed in metallic or non-metallic conduit or trunking that provides fire protection in accordance with BS EN 61386 series or BS EN 50085 series. This is to provide a degree of protection of at least IP4X.

11.8.2 Electrical connections

711.526.1 Joints should not be made in cables, except as a connection into a circuit.

Conductor connections should be within an enclosure providing a degree of protection of at least IP4X or IPXXD and the enclosure should incorporate a suitable cable anchorage.

11.8.3 Lighting installations
ELV lighting systems for filament lamps
Extra-low voltage systems for filament lamps must comply with BS EN 60598-2-23.   711.559.4.2

Lampholders
Insulation-piercing lampholders should not be used unless the cables and lampholders   711.559.4.3
are compatible and the lampholders are non-removable once fitted to the cable.

Electric discharge lamp installations
Installations of any luminous tube, sign or lamp as an illuminated unit on a stand, or   711.559.4.4
as an exhibit, with nominal power supply voltage higher than 230/400 V a.c., must
comply with Regulations 711.559.4.5 to 711.559.4.7.

The sign or lamp should be installed out of arm's reach or be adequately protected to
reduce the risk of injury to persons.

The facia or stand fitting material behind luminous tubes, signs or lamps must be
non-ignitable.

Emergency switching devices
A separate circuit should be used to supply signs, lamps or exhibits, which is controlled   711.559.4.7
by an emergency switch. The switch must be easily visible, accessible and clearly
marked.

Protection against thermal effects
Luminaires mounted below 2.5 m (arm's reach) from floor level or otherwise accessible   711.559.5
to accidental contact must be firmly and adequately fixed, and so sited or guarded as
to prevent risk of injury to persons or ignition of materials.

Note: In the case of outdoor lighting installations, Section 559 also applies, and a
degree of protection of at least IP33 may be required.

11.8.4 Electric motors
Where an electric motor might give rise to a hazard the motor must be provided with   711.55.4.1
an effective means of isolation on all poles and where necessary an emergency stop,   537.4.2.1
and the means should be adjacent to the motor it controls (see BS EN 60204-1).

11.8.5 Extra-low voltage transformers and electronic convertors
A manual reset protective device should protect the secondary circuit of each   711.55.6
transformer or electronic convertor.

Particular care should be taken to install ELV transformers out of reach of the public and
adequate ventilation should be allowed for. Access by a competent person for testing
and by a skilled person competent in such work for maintenance shall be provided.

Electronic convertors should conform with BS EN 61347-1.

11.8.6 Socket-outlets
An adequate number of socket-outlets should be installed to allow user requirements   711.55.7
to be met safely.

Floor mounted sockets should preferably not be installed, but where their use is
unavoidable they should be adequately protected from mechanical damage and
ingress of water.

11.9 Generators

Section 551 — Installations incorporating generator sets must comply with Section 551 of BS 7671. Where a generator is used to supply the temporary installation using a TN or TT system, it must be ensured that the installation is earthed, preferably by separate earth electrodes. For TN systems all exposed-conductive-parts should be bonded back to the generator. The neutral conductor and/or star point of the generator should be connected to the exposed-conductive-parts of the generator and reference earthed.

Part VI of the ESQC Regulations provides requirements for generation. Regulation 21 has requirements for switched alternative sources of energy (see Figure 11.1) as follows:

Where a person operates a source of energy as a switched alternative to a distributor's network, he shall ensure that that source of energy cannot operate in parallel with that network and where the source of energy is part of a low voltage consumer's installation, that installation shall comply with British Standard Requirements [meaning BS 7671].

The requirements for parallel operation are much more onerous.

11.10 Safety services

Chapter 56 — Where an exhibition is held within a building, it is assumed that the emergency lighting and/or fire safety systems etc. will be part of the permanent installation within that building. Care should be taken to ensure that existing emergency escape signs and escape routes are not obscured, impeded or blocked.

Additional emergency lighting should be installed in those areas not covered by the permanent installation. Where an exhibition is constructed out-of-doors, an adequate fire alarm system should be installed to enclosed areas to facilitate emergency evacuation.

Where an event is taking place out-of-doors and is open to the public in partial or total darkness, then:

▶ emergency lighting should be provided to escape routes
▶ provision should be made to ensure that alternative sources of supply for general lighting sufficient for safe evacuation are available throughout the area.

Reference should be made to the following British Standards:

▶ BS 5266 *Emergency lighting*
▶ BS 5839 *Fire detection and fire alarm systems for buildings.*

11.11 Inspection and testing

711.6 — The temporary electrical installations should be re-tested on site in accordance with Chapter 61 of BS 7671, after each assembly on site.

Users (e.g. exhibitors and stall holders) should be advised to visually check electrical equipment for damage on a daily basis.

Floor and ceiling heating systems

12

12.1 Introduction

This is a new section that has been introduced into the 17th Edition.

Section 753

Regulation 753.411.3.2 has been included so that 30 mA RCDs are required for ceiling or floor heating systems to protect against penetration of the heating elements.

There are requirements to protect the building structure against fire and to avoid the overheating of the heating system.

Heating units manufactured without exposed-conductive-parts must be provided on site with a metal grid.

12.2 Scope

Section 753 applies to the installation of electric floor and ceiling heating systems which are erected as either thermal storage heating systems or direct heating systems. It does not apply to the installation of wall heating systems. Heating systems for use outdoors are not considered.

753.1

Note: A ceiling located under the roof of a building down to a vertical height of 1.50 m measured from the finished floor surface is also regarded as a ceiling within the meaning of the Regulations.

12.3 The risks

The risks associated with ceiling heating systems are penetration of the heating elements by nails, drawing pins, etc. pushed through the ceiling surface. For this reason, additional protection is required by the use of a 30 mA RCD.

753.415.1

Similarly, there are concerns that underfloor heating installations can be damaged by carpet gripper nails, etc. and for similar reasons protection by a 30 mA RCD is required.

There is a risk of fire to the building structure due to the use of heating, therefore there are requirements to avoid overheating of the floor or ceiling heating system.

12.4 Protection against electric shock

753.410   The protective measures of obstacles, placing out of reach, non-conducting location and earth-free local equipotential bonding are not permitted.

12.4.1 Protective measure: automatic disconnection of supply

753.411.3.2   RCDs with a rated residual operating current not exceeding 30 mA shall be used as disconnecting devices. For heating units which are delivered from the manufacturer without exposed-conductive-parts, a suitable conductive covering, e.g. a grid with a spacing of not more than 30 mm, must be provided on site as an exposed-conductive-part above the floor heating elements or under the ceiling heating elements, and connected to the protective conductor of the electrical installation.

Note: Limitation of the rated heating power to 7.5 kW/230 V or 13 kW/400 V downstream of a 30 mA RCD may avoid unwanted tripping due to leakage capacitance. Values of leakage capacitance may be obtained from the manufacturer of the heating system.

12.5 Protection against overheating

753.424.1.1   To avoid overheating of floor or ceiling heating systems in buildings, at least one of the following measures shall be applied to limit the temperature and the heating zone to a maximum of 80 °C:

i   Appropriate design of the heating system
ii   Appropriate installation of the heating system
iii   Use of protective devices.

Heating units shall be connected to the electrical installation via cold tails or suitable terminal fittings.

Heating units shall be inseparably connected to the cold tails, e.g. by welding, brazing or by compression jointing techniques.

Heating units must not cross expansion joints.

GN4   Additional guidance for protection against fire for this type of location can be found in section 8.5 of Guidance Note 4: *Protection Against Fire*.

753.424.1.2   As the heating unit may cause higher temperatures or arcs under fault conditions, special measures to meet the requirements of Chapter 42 should be taken when the heating unit is installed close to easily ignitable building structures, such as placing on a metal sheet, in metal conduit or at a distance of at least 10 mm in air from the ignitable structure.

12.6 Standards

753.511   Flexible sheet heating elements should comply with BS EN 60335-2-96 and heating cables should comply with BS 6351 series.

12.7 External influences

Heating units for installation in ceilings shall be at least IPX1 and heating units for installation in floors of concrete or similar material shall be not less than IPX7 with appropriate mechanical properties.

753.512.2

12.7.1 Operational conditions

Precautions shall be taken not to stress the heating unit mechanically; for example, the material by which it is to be protected in the finished installation shall cover the heating unit as soon as possible.

753.512.1

12.8 Identification

The designer of the installation/heating system or installer shall provide a plan for each heating system, containing the following details:

753.514

1 Manufacturer and type of heating units
2 Number of heating units installed
3 Length/area of heating units
4 Rated power
5 Surface power density
6 Layout of the heating units in the form of a sketch, a drawing or a picture
7 Position/depth of heating units
8 Position of junction boxes
9 Conductors, shields and the like
10 Heated area
11 Rated voltage
12 Rated resistance (cold) of the heating units
13 Rated current of overcurrent protective devices
14 Rated residual operating current of RCD
15 The insulation resistance of the heating installation and the test voltage used
16 The leakage capacitance.

The plan shall be fixed to, or adjacent to, the distribution board of the heating system.

Section 12.11 describes advice to be provided by the installer for the user of the installation.

12.9 Heating-free areas

Where heating units are installed, there shall be heating-free areas where drilling and fixing by screws, nails and the like are permitted. The installer shall inform other contractors that no penetrating means, such as screws for door stoppers, shall be used in the area where floor or ceiling heating units are installed.

753.522.4

It may be necessary to provide areas of floor or ceiling that are unheated, e.g. where fixtures to the floor or ceiling would prevent the proper emission of heat.

753.520.4

Account shall be taken of the increase in ambient temperature and of its effect upon the cables, including cold tails (circuit wiring) and control wiring installed in heated zones.

753.522.1.3

12.10 Locations containing a bath or shower and swimming pools and other basins

701.753
702.55.1
Where underfloor heating systems are installed in bathrooms and swimming pools supplied at voltages other than SELV, then the heating element should be provided either with a metallic sheath or overall screen or a metallic grid positioned above the heating elements. The screen or grid shall be connected to the supplementary bonding and earthing for the facility. In addition, the supply to the heating elements should be controlled via an RCD with a rated residual operating current not exceeding 30 mA with no adjustable time delay.

As such facilities are likely to suffer from the ingress of water via tiling etc. and also from corrosion, provision for ready access to all terminations to permit the thorough testing of such installations is vital. The use of suitable plastic materials for enclosures and conduit/trunking etc. is one method of minimising potential corrosion effects. Furthermore, regular visual inspection of these installations, particularly in swimming pools where corrosion from the chemicals is more likely, should be carried out and adequate records kept.

Sections 701/702
All such facilities must comply with, as applicable, Sections 701 and/or 702 of BS 7671:2008.

12.11 Information from the contractor for the user of the installation

Fig 753 of BS 7671
A description of the heating system shall be provided by the installer/contractor of the heating system to the owner of the building or his/her agent upon completion of the installation.

The description shall contain at least the following information:

a Description of the construction of the heating system, which must include the installation depth of the heating units;
b Location diagram with information concerning
  ▶ the distribution of the heating circuits and their rated power;
  ▶ the position of the heating units in each room;
  ▶ conditions which have been taken into account when installing the heating units, e.g. heating-free areas, complementary heating zones, unheated areas for fixing means penetrating into the floor covering;
c Data on the control equipment used, with relevant circuit diagrams and the dimensioned position of floor temperature and weather conditions sensors, if any;
d Data on the type of heating units and their maximum operating temperature.

The installer/contractor shall inform the owner that the description of the heating system includes all necessary information, e.g. for repair work.

The designer/installer/contractor of the heating system shall hand over an appropriate number of instructions for use to the owner or his/her agent upon completion. One copy of the instructions for use shall be permanently fixed in or near each relevant distribution board.

The instructions for use shall include at least the following information:

a Description of the heating system and its function;

b Operation of the heating installation in the first heating period in the case of a new building, e.g. regarding drying out;

c Operation of the control equipment for the heating system in the dwelling area and the complementary heating zones as well, if any;

d Information on restrictions on placing of furniture or similar. Information provided to the owner shall cover the restrictions, if any, including:

  ▶ whether additional floor coverings are permitted, for example, carpets with a thickness of >10 mm may lead to higher floor temperatures which can adversely affect the performance of the heating system

  ▶ where pieces of furniture solidly covering the floor and/or built-in cupboards may be placed on heating-free areas

  ▶ where furniture, such as carpets, seating and rest furniture with pelmets, which in part do not solidly cover the floor, may not be placed in complementary heating zones, if any;

e Information on restrictions on placing of furniture or similar;

f In the case of ceiling heating systems, restrictions regarding the height of furniture. Cupboards of room height may be placed only below the area of ceiling where no heating elements are installed;

g Dimensioned position of complementary heating zones and placing areas;

h Statement that, in the case of thermal floor and ceiling heating systems, no fixing shall be made into the floor and ceiling respectively. Excluded from this requirement are unheated areas. Alternatives shall be given, where applicable.

Gardens (other than horticultural installations)

13

13.1 Introduction

Gardens and similar installations are not classified as special locations in BS 7671. However, electrical installation work in these locations is notifiable under Part P (Electrical safety – dwellings) of the Building Regulations 2000. Specifically this applies in a garden or in or on land associated with a building where the electricity is from a source located within or shared with a dwelling.

The general rules for outdoor circuits and equipment apply to all gardens.

13.2 The risks

Gardens can present a number of potential risks where electrical installations are concerned. These risks may include:

▶ contact of persons with the general mass of Earth, possibly with bare feet
▶ a frequently wet environment
▶ the wearing of minimal clothing
▶ gardening activity that may cause damage or disturbance to cables or equipment
▶ insertion of spikes in the ground for securing marquees, inflatables etc.

In public gardens it should be recognised that electrical equipment, e.g. lighting, may be accessible to the general public.

13.3 Buried cables

Cables should be protected against foreseeable damage, either by armouring or by suitable enclosure. Unprotected cables should not be buried direct in the ground nor should they be clipped to wooden fences, etc. which may provide inadequate support and protection.

Problems arise when either ground levels are lowered so that cables have insufficient cover or when ground levels are raised so that cables which were not intended to be buried and are not suitable for burial become buried. Such problems can arise during the course of a project and the intended ground level should be formally ascertained before the cables are installed. It must be remembered that the layout of a garden can be changed totally within a few seasons and great care must be taken to route cables where they are not likely to be disturbed or damaged, e.g. around the edge of the plot and at sufficient depth.

522.8.10 Buried cable routes should be identified by local route markers, and recorded on drawings; cables should be buried at least 450 mm (preferably deeper) below the lowest local ground level, and a route marker tape laid along the cable route about
522.2.1 150 mm below the surface. Cables should be shielded from prolonged exposure to
522.11.1 direct sunlight, or be of a type suitable for such exposure. Cable with a black sheath is recommended. Generally, the ultraviolet light from the sun will affect plastics and the cable manufacturer's advice should be taken. However, cables must not be so
522.8.4 enclosed that heat dissipation is inadequate. Cables taken overhead should have a suitable rigid support or a catenary wire, or be a type suitable for such installation. See
GN1 Guidance Note 1, Appendix G.3 for more information.

13.4 Socket-outlets

411.3.3 Mobile equipment with a current rating not exceeding 32 A for use outdoors requires additional protection by means of a 30 mA RCD with characteristics in accordance with Regulation 415.1. Socket-outlets that can be expected to be used to supply this equipment outdoors will therefore require this additional protection, which can be provided by using an RCCB complying with BS 61008-1 or BS 4293, an RCBO complying with BS EN 61009-1 or an SRCD complying with BS 7288.

Socket-outlets should be suitably placed to be convenient for the purpose, and of IP rating e.g. IP44 or better if located outdoors.

411.3.3 note 2 An RCD is not required for a socket-outlet in this instance if the circuit is reduced low voltage or extra-low voltage.

13.5 Fixed equipment

Fixed equipment in the garden, such as permanently fixed garden lighting, should be securely fixed with all cables buried, or supported away from the ground or paths. All-insulated Class II equipment is recommended where possible for increased safety. Decorative lighting, including 'festoon' type lighting, should be permanently fixed if
Section 559 used regularly. See Section 559 of BS 7671:2008.

Other fixed equipment, including pumps, etc., should be installed in accordance with the manufacturer's instructions and the general requirements of BS 7671:2008.

13.6 Ponds

Section 702 In view of the risk of accidental or intentional immersion it is recommended that the same rules should be applied to garden ponds, especially larger ones, as are applied to swimming pools. Equipment (including cables) must be suitable for the purpose and of a suitable IP rating, or be installed in a suitable enclosure. Class II equipment should be utilised where possible. Cables should be installed in ducts or conduits built into the pond structure and not allowed to lie loose around the area. All connections must be made in robust, watertight junction boxes. Equipment to IP55 or better is recommended.

Pond lighting should meet the requirements of BS EN 60598-2-18, pumps BS EN 60335-2-41, and other equipment BS EN 60335-2-55.

Mobile or transportable units

14.1 Introduction

The 16th Edition did not include requirements for mobile or transportable units, but guidance was given in the previous edition of this Guidance Note. However, new Section 717 concerning the electrical installations of mobile or transportable units has been introduced into the 17th Edition.

Section 717 specifies particular requirements for this type of electrical installation relating to protection against electric shock and selection and erection of equipment.

Section 717

14.2 Scope

The particular requirements of Section 717 are applicable to mobile or transportable units. For the purposes of this section, the term 'unit' is intended to mean a vehicle and/or mobile or transportable structure in which all or part of an electrical installation is contained, which is provided with a temporary supply by means of, for example, a plug and socket-outlet.

Units are either:

i of the mobile type, e.g. vehicles (self-propelled or towed), or
ii of the transportable type, e.g. containers or cabins.

Examples of the units include technical and facilities vehicles for the entertainment industry, medical services, advertising, fire fighting, workshops, offices, transportable catering units etc. The requirements are not applicable to:

i generating sets
ii marinas and pleasure craft
iii mobile machinery in accordance with BS EN 60204-1
iv caravans to Section 721
v traction equipment of electric vehicles
vi electrical equipment required by a vehicle to allow it to be driven safely or used on the highway.

The guidance given here is in addition to the general requirements of BS 7671. Where other special locations such as rooms containing showers or medical locations form part of a mobile or transportable unit, the special requirements for those installations should also be taken into consideration. In this regard, particular reference should be made to Part 7 of BS 7671 and the other relevant chapters of this Guidance Note.

14.3 The risks

The risks associated with mobile and transportable units arise from:

i Risk of loss of connection to earth, due to use of temporary cable connections and long supply cable runs; the repeated use of cable connectors which may give rise to 'wear and tear' and the potential for mechanical damage to these parts.

ii Risks arising from the connection to different national and local electricity distribution networks, where unfamiliar supply characteristics and earthing arrangements are found.

iii Impracticality of establishing an equipotential zone external to the unit.

iv Open-circuit faults of the PEN conductor of PME supplies raising the potential of all metalwork (including that of the unit) to dangerous levels.

v Risk of shock arising from high functional currents flowing in protective conductors – usually where the unit contains substantial amounts of electronics or communications equipment.

vi Vibration while the vehicle or trailer is in motion, or while a transportable unit is being moved – causing faults within the unit installation.

Particular requirements to reduce these risks include:

a Checking the suitability of the electricity supply before connecting the unit.

b Accessible conductive parts of the unit to be connected through the main equipotential bonding to the main earthing terminal within the unit.

c A regime of regular inspection and testing of connecting cables and their couplers, supported by a log-book system of record keeping.

d The use of flexible cables.

e The use of RCDs.

f Clear and unambiguous labelling of units, indicating types of supply which may be connected.

g Particular attention paid to the maintenance and periodic inspection of installations.

14.4 Supplies

14.4.1 General
Figures 14.1 to 14.4 show examples of types of supply as follows:

Fig 717.1 **14.1** Example of connection to low voltage generating set located inside the unit, with or without an earth electrode,

Fig 717.2 **14.2** Example of connection to a low voltage generating set located outside the unit,

Fig 717.3 **14.3** Example of connection to a TN or TT electrical installation, with or without an earth electrode at the unit,

Fig 717.4 **14.4** Example of connection to a fixed electrical installation with any type of earthing system (using simple separation and an internal TN system, with or without an earth electrode).

717.411.4 A TN-C-S system shall not be used to supply a mobile or transportable unit except:

i where the installation is continuously under the supervision of a skilled or instructed person, and

ii the suitability and effectiveness of the means of earthing has been confirmed before the connection is made.

▼ **Figure 14.1** Example of connection to low voltage generating set located inside the unit, with or without an earth electrode

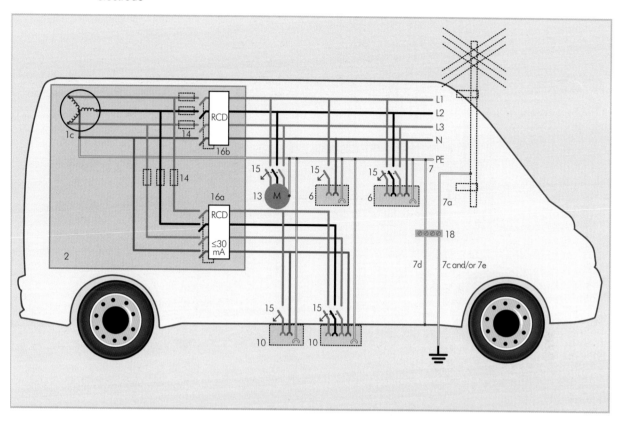

▼ **Figure 14.2** Example of connection to a low voltage generating set located outside the unit

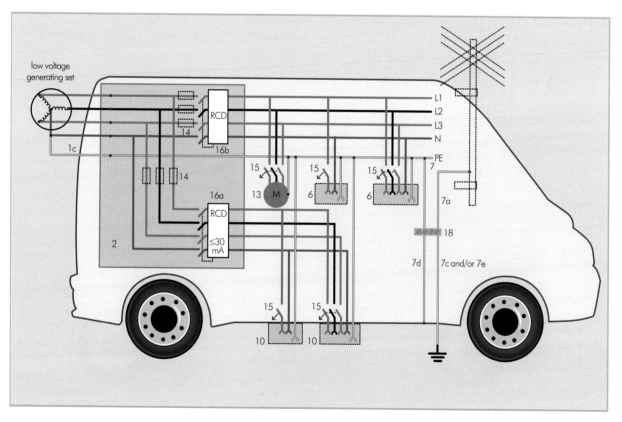

▼ **Figure 14.3** Example of connection to a TN or TT electrical installation, with or without an earth electrode at the unit

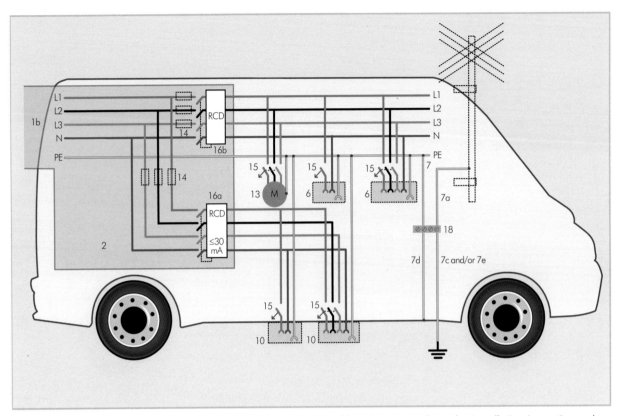

Note: TN-C-S should not be used to supply a mobile or transportable unit except where the installation is continuously under the supervision of skilled or instructed persons in accordance with Regulation 717.411.4.

▼ **Figure 14.4** Example of connection to a fixed electrical installation with any type of earthing system (using simple separation and an internal TN system, with or without an earth electrode)

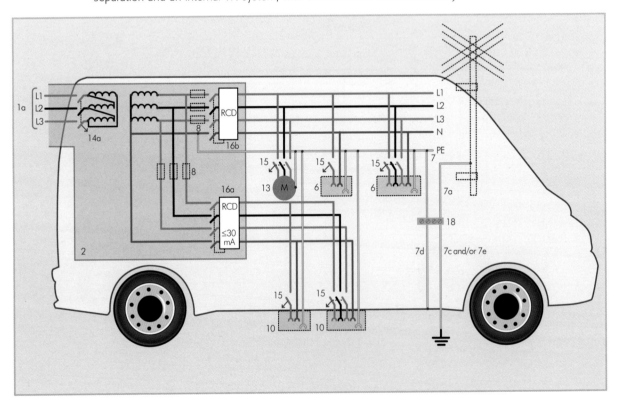

Key to Figures 14.1 to 14.4:

1a Connection of the unit to a supply through a transformer with simple separation

1b Connection of the unit to a supply in which the protective measures are effective

1c Connection to an LV generator set in accordance with Section 551

2 Class II or equivalent enclosure up to the first protective device providing automatic disconnection of supply

6 Socket-outlets for use exclusively within the unit

7 Main equipotential bonding in accordance with Regulation 717.411.3.1.2

7a to an antenna pole, if any

7c to a functional earth electrode, if required

7d to the conductive enclosure of the unit

7e to an earth electrode for protective purposes, if required

8 Protective devices, if required, for overcurrent and/or for protection by disconnection of supply in case of a second fault

10 Socket-outlets for current-using equipment for use outside the unit

13 Current-using equipment for use exclusively within the unit

14 Overcurrent protective device, if required

14a Overcurrent protective device, if required

15 Overcurrent protective device

16a RCD having the characteristics specified in Regulation 415.1.1 for protection by automatic disconnection of supply for circuits of equipment for use outside the unit

16b RCD for protection by automatic disconnection of supply for circuits of equipment for use inside the unit

18 Main earthing terminal or bar.

14.4.2 IT systems

An IT system can be provided by: 711.411.6.2

a an isolating transformer or a low voltage generating set, with an insulation monitoring device installed, or

b a transformer providing simple separation, e.g. in accordance with BS EN 61558-1, only in the following cases:

 ▶ an insulation monitoring device is installed with or without an earth electrode, providing automatic disconnection of the supply in case of a first fault between live parts and the frame of the unit, or

 ▶ an RCD and an earth electrode are installed to provide automatic disconnection in the case of failure in the transformer providing the simple separation. Each item of equipment used outside the unit shall be protected by a separate RCD having the characteristics specified in Regulation 415.1.1.

14.5 Protection against electric shock

717.417 The protective measures of obstacles and placing out of reach (Section 417) are not permitted.

717.418 The protective measure of non-conducting location (Regulation 418.1) is not permitted and earth-free local equipotential bonding (Regulation 418.2) is not recommended.

717.415 Every socket-outlet intended to supply current-using equipment outside the unit shall be provided with additional protection by an RCD with a rated residual operating current not exceeding 30 mA, with the exception of socket-outlets which are supplied from circuits with protection by:

i SELV, or
ii PELV, or
iii electrical separation.

14.6 Protective equipotential bonding

717.411.3.1.2 Accessible conductive parts of the unit, such as the chassis, shall be connected through the main protective bonding conductors to the main earthing terminal within the unit.

The protective bonding conductors shall be finely stranded. Cable types H05V-K and H07V-K to BS 6004 are considered appropriate.

14.7 Wiring systems external to the mobile or transportable unit

717.52.1 Flexible cables (for connecting the unit to the supply) in accordance with H07RN-F (BS 7919), or cables of equivalent design, having a minimum cross-sectional area of 2.5 mm² copper shall be used.

The flexible cable shall enter the unit by an insulating inlet in such a way as to minimise the possibility of any insulation damage or fault which might energize the exposed-conductive-parts of the unit.

14.8 Wiring systems inside the mobile or transportable unit

717.52.2 The following or other equivalent cable types are permitted for the internal wiring of the unit:

i Thermoplastic or thermosetting insulated only cable (BS 6004, BS 7211, BS 7919) installed in conduits in accordance with BS EN 61386-1
ii Thermoplastic or thermosetting insulated and sheathed cable (BS 6004, BS 7211, BS 7919), if precautionary measures are taken to prevent mechanical damage due to any sharp-edged parts or abrasion.

14.9 Proximity to non-electrical services

No electrical equipment including wiring systems, except ELV equipment for gas supply control, shall be installed in any gas cylinder storage compartment.

717.528.3

Where cables have to run through such a compartment, they shall be run at a height of less than 500 mm above the base of the cylinder(s), and such cables shall be protected against mechanical damage by installation within a continuous gas-tight conduit or duct passing through the compartment. Where installed, this conduit or duct shall be able to withstand an impact equivalent to AG3 without visible physical damage.

Appx 5

14.10 Identification and labelling

A permanent notice of durable material shall be fixed to the unit in a prominent position, preferably adjacent to the supply inlet connector. The notice should state in clear and unambiguous terms the following information:

717.514

- ▶ The type of supply which may be connected to the unit.
- ▶ The voltage rating of the unit.
- ▶ The number of phases and their configuration.
- ▶ The on-board earthing arrangement.
- ▶ The maximum power requirement of the unit.

14.11 Other equipment

Plugs and connectors used to connect the unit to the supply shall comply with BS EN 60309-2 and shall also meet the following requirements:

717.55.1

i   Plugs shall have an enclosure of insulating material
ii Plugs and socket-outlets shall afford a degree of protection of not less than IP44, if located outside
iii Appliance inlets with their enclosures shall provide a degree of protection of at least IP44
iv The plug part shall be situated on the unit.

Socket-outlets located outside the unit shall be provided with an enclosure affording a degree of protection not less than IP44.

717.55.2

14.12 Routine maintenance and testing

The service duty of mobile and transportable units will vary with the type of unit and type of use (e.g. owner-operator or hire) but it is likely that frequent connecting and disconnecting combined with transporting will amount to the equivalent of rough service life. Frequency of use should therefore be an important factor in determining inspection and testing intervals. It is recommended that a visual inspection is carried out on the connecting cable and all plugs and socket-outlets before each and every transported use of the unit. The results of the visual inspection should be entered in a log-book as a permanent record of the condition of the electrical equipment. No repairs or extensions are acceptable on the external cable system and this should be replaced in its entirety if there are signs of damage or wear-and-tear.

As a minimum, the unit electrical system should be inspected and tested annually, a report obtained on its condition and the necessary maintenance work, if any, implemented before the unit is next used. The recommendations of Part 6 of BS 7671 should be followed in this regard, together with the specific guidance given in IEE

GN3   Guidance Note 3. If the unit duty is considered to be arduous, the inspection intervals should be reduced to cater for the particular conditions experienced.

All RCDs should be tested regularly by operating the test button and periodically by a proprietary test instrument to ensure they conform to the parameters of the relevant British Standard.

Appx 6   All tests should be tabulated for record purposes. The necessary forms required by Part 6 of BS 7671 must be provided to the person ordering the work, by the contractor or persons carrying out the inspection and tests.

Small-scale embedded generators (SSEG)

15

15.1 Introduction and the law

Low voltage generating sets such as small-scale embedded generators are not classed as special locations or installations under BS 7671:2008. There are, however, regulations related to this type of installation in Section 551 'Low voltage generating sets'. In addition, BS EN 50438:2007 gives the requirements for the connection of micro-generators in parallel with public low voltage distribution networks.

Section 551

For domestic installations small-scale embedded generators such as micro-CHP units are classified as special installations under Approved Document P (Electrical safety – dwellings) of the Building Regulations 2000.

15.1.1 The Electricity Safety, Quality and Continuity Regulations 2002

Within the lifetime of this edition of Guidance Note 7 the use of small-scale embedded generators (SSEG) is likely to become widespread. The Electricity Safety, Quality and Continuity Regulations 2002 exempt sources of energy with an electrical output not exceeding 16 A per phase at low voltage (230 V) from Regulations 22(1)(b) and 22(1)(d) of the Regulations, and this chapter covers such sources of energy.

In addition to the electrical output limit of 16 A per phase (sub-paragraph (a) of Regulation 22(2)), other requirements for such small generators are:

> (b) the source of energy is configured to disconnect itself electrically from the parallel connection when the distributor's equipment disconnects the supply of electricity to the consumer's installation; and
>
> (c) the person installing the source of energy ensures that the distributor is advised of the intention to use the source of energy in parallel with the network before, or at the time of, commissioning the source.

The requirements that still need to be met in Regulation 22(1) are:

> ... No person shall install or operate a source of energy which may be connected in parallel with a distributor's network unless he –
>
> (a) has the necessary and appropriate equipment to prevent danger or interference with that network or with the supply to consumers so far as is reasonably practicable;
>
> (c) where the source of energy is part of a low voltage consumer's installation, complies with the provisions of British Standard Requirements [meaning, by definition, BS 7671].

15.1.2 BS 7671:2008 Section 551 – Low voltage generating sets

551.2 Section 551 now includes additional requirements contained in Regulation 551.2 to ensure the safe connection of low voltage generating sets including small-scale embedded generators.

551.4.2 A new Regulation 551.4.2, regarding the use of RCDs, has been added. This regulation states:

> *The generating set shall be connected so that any provision within the installation for protection by RCDs in accordance with Chapter 41 remains effective for every intended combination of sources of supply.*

551.1 Note Notes have been added including one to Regulation 551.1 stating that the procedure for connecting generating sets up to 16 A in parallel with the public supply is given in ESQCR 2002. Requirements of the distributor for the connection of units rated up to 16 A are given in BS EN 50438 *Requirements for the connection of micro-generators in parallel with public low-voltage distribution networks.*

551.7.2 For sets above 16 A the requirements of the distributor must be ascertained. The 17th Edition recognises that there are two connection options:

i connection into a separate dedicated circuit
ii connection into an existing final circuit.

Connection into a dedicated circuit is preferred.

Regulation 551.7.2 sets out the requirements for the two options. The regulation requires that a generating set used as an additional source of supply in parallel with another source shall either be installed on the supply side of all protective devices for the final circuits of the installation (connection into a separate dedicated circuit) or, if connected on the load side of all protective devices for the final circuits, must fulfil a number of additional requirements. These additional requirements are:

i The current-carrying capacity of the final circuit conductors shall be greater than or equal to the rated current of the protective device plus the rated output of the generating set
ii A generating set shall not be connected to a final circuit by a plug and socket
iii A residual current device providing additional protection of the final circuit in accordance with Regulation 415.1.1 shall disconnect all live conductors including the neutral conductor
iv The line and neutral conductors of the final circuit and of the generating set shall not be connected to Earth
v Unless the device providing automatic disconnection of the final circuit in accordance with Regulation 411.3.2 disconnects the line and neutral conductors, it shall be verified that the combination of the disconnection time of the protective device for the final circuit and the time taken for the output voltage of the generating set to reduce to 50 V or less is not greater than the disconnection time required by Regulation 411.3.2 for the final circuit.

The generator equipment should be type-tested and approved by a recognised body.

15.2 Engineering Recommendation G83

To assist network operators and installers the Energy Networks Association (formerly the Electricity Association) has prepared Engineering Recommendation G83: *Recommendations for the connection of small-scale embedded generators (up to 16 A per phase at low voltage) in parallel with public low voltage distribution networks*. The guidance given in this chapter is intended to reproduce the requirements of the Engineering Recommendation as they would apply to persons responsible for electrically connecting such generators.

The Engineering Recommendation is for all small-scale embedded generator (SSEG) installations with an output up to and including 16 A per phase, single or multiphase, 230/400 V a.c. This includes:

▶ domestic combined heat and power
▶ hydro
▶ wind power
▶ photovoltaic
▶ fuel cells.

Requirements for solar photovoltaic installations are found in Chapter 10 of this Guidance Note (see also section 15.7.2).

The Engineering Recommendation describes two types of connection procedure for the connection of an SSEG unit(s) in parallel with the public low voltage distribution network.

The first is for a single SSEG unit referred to as a Stage 1 connection within a single customer's installation. The installer first has to make the distribution network operator (DNO) aware of the SSEG installation at or before the time of commissioning in accordance with the ESQC Regulation 22(2)(c). The installer then has to provide the DNO with all the necessary information on the installation within 30 days of the SSEG unit being commissioned.

The second is for a planned installation project referred to as a Stage 2 connection where there is a proposal to install multiple SSEG units in a close geographical region. The installer would be expected to discuss the project with the local DNO prior to installation who will assess the impact of the connections on the network and provide the appropriate connection conditions. The installer then has to provide the DNO with all the necessary information on the installation within 30 days of commissioning.

Engineering Recommendation G83 incorporates forms, which define the information required by a public distribution network operator for a small-scale embedded generator which is connected in parallel with a public low voltage distribution network. Supply of information in this form, for a suitably type-tested unit, is intended to satisfy the legal requirements of the DNO and hence will satisfy the legal requirements of the Electricity Safety, Quality and Continuity Regulations 2002.

15.3 Installation and wiring

The installation that connects the embedded generator to the supply terminals must comply with BS 7671.

A suitably rated overcurrent protective device shall protect the wiring between the electricity supply terminals and the embedded generator.

The SSEG shall be connected directly to a local isolating switch. For single-phase machines the phase and neutral are to be isolated and for multiphase machines all phases and neutral are to be isolated. In all instances the switch, which must be manual, shall be capable of being secured in the 'off' isolating position. The switch is to be located in an accessible position in the customer's installation.

15.4 Means of isolation

551.7.6   Regulation 551.7.6 states the requirements for isolation:

> **551.7.6** Means shall be provided to enable the generating set to be isolated from the system for distribution of electricity to the public. For a generating set with an output exceeding 16 A, the accessibility of this means of isolation shall comply with national rules and distribution system operator requirements. For a generating set with an output not exceeding 16 A, the accessibility of this means of isolation shall comply with BS EN 50438.

15.5 Earthing

When a SSEG is operating in parallel with a distributor's network, there shall be no direct connection between the generator winding (or pole of the primary energy source in the case of PV array or fuel cell) and the network operator's earth terminal. See Figure 15.1.

▼ **Figure 15.1**
Earthing of parallel operation SSEG

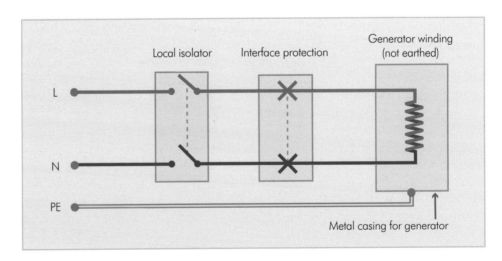

For all connections earthing arrangements shall comply with the requirements of BS 7671.

15.6 Labelling

Labels (i.e. warning notices) are required at:

514.15.1

▶ the supply terminals (fused cut-out),
▶ the meter position,
▶ the consumer unit, and
▶ all points of isolation of both sources of supply to indicate the presence of the SSEG within the premises.

The Health and Safety (Safety Signs and Signals) Regulations 1996 stipulate that the labels should display the prescribed triangular shape and size using black on yellow colouring. A typical label both for size and content is shown in Figure 15.2.

▼ **Figure 15.2**
Warning notice for dual supply

In addition, Engineering Recommendation G83 requires up-to-date information to be displayed at the point of connection with a distributor's network as follows:

a A circuit diagram showing the circuit wiring, including all protective devices, between the embedded generator and the network operator's fused cut-out. This diagram is also required to show by whom the generator is owned and maintained

b A summary of the protection settings incorporated within the equipment.

Figure 15.3 is an example of the type of circuit diagram that needs to be displayed. This diagram is for illustrative purposes and not intended to be fully descriptive. Metering devices, consumer unit or circuit protective devices within the installation would need to be shown.

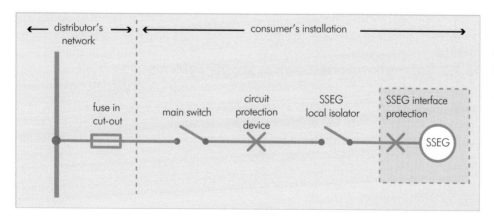

▼ **Figure 15.3**
Example of circuit diagram for an SSEG installation (from Engineering Recommendation G83)

The installer is required to advise that it is the user's responsibility to ensure that this safety information is kept up to date. The installation operating instructions shall contain the SSEG manufacturer's contact details, e.g. name, telephone number and web address.

15.7 Types of SSEG

Note: SSEGs will only operate in parallel with the mains supply.

15.7.1 Domestic combined heat and power

Annex B of Engineering Recommendation G83 specifies the particular requirements for combined heating and power sets. These are likely to be the most common type of set encountered by the electricity installer. They can be incorporated into a household gas boiler to generate electricity.

Most small combined heat and power generators embody a Stirling engine – see Figure 15.4.

▼ **Figure 15.4**
The Stirling engine

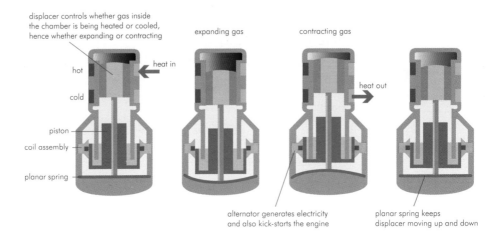

The Stirling engine does not burn the gas within the cylinder as in an internal combustion engine. The power to the engine is delivered by the combustion gases from the gas burner, and the energy transfer is in effect achieved by the temperature difference between the burner exhaust gases and burner input air or circulating water.

The basic principle is that the sealed gas within the engine is heated by the burner gases and expands. On expansion the driven piston generates electricity in the winding and compresses the planar spring. In the expanded, say, down position of the piston the gas within the piston is subject to the cooling effect of the cooler input air (or circulated water) and contracts, assisted by the planar spring. At the compression position the gas in the cylinder is now heated by the burner gases only and is not subject to the cooling effects of the input air.

15.7.2 Solar photovoltaic power supply systems

Chapter 10 of this publication gives guidance for solar photovoltaic power supply systems. The requirements are for power photovoltaic systems that will generate only when run in parallel with the electricity supply. Requirements for photovoltaic power supply systems which are intended for stand-alone operation are under consideration by the IEC and not considered here.

Temporary electrical installations for structures, amusement devices and booths at fairgrounds, amusement parks and circuses

16

16.1 Introduction

This is a new section that has been introduced into the 17th Edition and it also appears in this Guidance Note for the first time.

Section 740

Section 740 recognises that some installations are exposed to many differing and onerous circumstances, as they are often frequently installed, dismantled, moved to a new location then installed and operated again.

To compound problems, such installations can be exposed to the elements, and the locations are open to the general public, may house animals and livestock, and are also a place of work.

The equipment must function without compromising safety; therefore, the installation has to be fit for purpose and be designed to cope with ever-changing conditions.

Section 740 prescribes particular measures to reduce the risks caused by this harsh treatment of the electrical installation.

16.2 Scope

Section 740 specifies the minimum electrical installation requirements to facilitate the safe design, installation and operation of temporarily erected mobile or transportable electrical machines and structures which incorporate electrical equipment. The machines and structures are intended to be installed repeatedly, without loss of safety, temporarily, at fairgrounds, amusement parks, circuses or similar places.

740.1.1

Note: The permanent electrical installation, if there is one, is excluded from the scope.

The object of this section is to define the electrical installation requirements for such structures and machines, being either integral parts or constituting the total amusement device. The scope does not include the internal electrical wiring of machines (see BS EN 60204-1).

16.3   Electrical supplies

740.313.1.1   The nominal supply voltage of temporary electrical installations in booths, stands and amusement devices should not exceed 230/400 V a.c. or 440 V d.c. Supplies can be obtained from a number of sources:

i   from the public network, i.e. the DNO
ii   generators, i.e. those mounted on trucks owned by the touring event
iii   from privately owned supplies, i.e. a local factory with sufficient spare capacity.

740.313.3   There can be any number of electrical sources supplying the temporary system and it is of paramount importance that line and neutral conductors from different sources are not interconnected.

Where the supply is obtained from the DNO any instructions given must be adhered to. Supplies obtained from the DNO would preferably be TN-S but this isn't always possible. A TN-S system has the neutral of the source of energy connected with Earth at one point only, at, or as near as is reasonably practicable to, the source of supply. The consumer's main earthing terminal is typically connected to the metallic sheath of the distributor's SWA service cable.

740.411.4   Where the available supply is TN-C-S, the supply should not be used in that form, i.e. a TT system should be created. The reason is that the ESQCR prohibit the use of a TN-C-S system for the supply to a caravan or similar construction.

740.411.6   Where continuity of service is important, IT systems may be used for d.c. applications only.

16.4   Protection against electric shock

740.410.3   At the origin of each electrical supply, to all or part of the installation, an RCD, with a rated residual operating current not exceeding 300 mA, is to be installed to provide automatic disconnection of supply. As there will be further RCDs downstream of this point, this RCD should be of the S-type complying with the requirements of BS EN 61008-1 or BS EN 61009-1, and incorporate a time delay in accordance with BS EN 60947-2, to provide discrimination with further RCDs protecting final circuits. For supplies to a.c. motors, RCDs should be the time-delayed or S-type where necessary to prevent unwanted tripping.

The protective measure of protection by obstacles is not permitted in this type of installation; however, placing out of arm's reach is acceptable for electric dodgems – see section 16.8.2.

740.410.3.6   The protective measures of non-conducting location (Regulation 418.1) and earth-free local equipotential bonding (Regulation 418.2) are not permitted.

16.5 Additional protection

All final circuits in the installation for lighting, socket-outlets rated up to 32 A, and mobile equipment connected by a flexible cable and rated up to 32 A, are to be protected by an RCD having a rated residual operating current not exceeding 30 mA.

740.415.1

The requirement for additional protection relates to the increased risk of damage to cables within an installation of this nature.

Lighting circuits incorporating emergency luminaires, with self-contained batteries for example, should be protected by the same RCD protecting that lighting circuit.

Additional protection is not required for:

i   SELV or PELV circuits – this measure alone is deemed to be a protective measure in all situations
ii circuits protected by electrical separation
iii lighting circuits placed out of arm's reach – provided they are not supplied by socket-outlets for household or similar purposes, e.g. those manufactured to BS 1363 or socket-outlets according to BS EN 60309-1. Typically, this would be achieved by devices for connection of luminaires to BS EN 61995 or luminaire supporting couplers to BS 6972.

To minimise potentials, supplementary bonding should be installed to connect all exposed-conductive-parts and extraneous-conductive-parts that can be touched by livestock. Where a metal grid is laid in the floor, or extraneous-conductive-parts are accessible, they should be included within the supplementary bonding of the location. It is important to note that animal excrement and urine are very corrosive and so all supplementary bonding connections should be protected by suitable enclosure.

740.415.2

16.6 Installation

16.6.1 Wiring systems

Conduit, cable trunking, cable ducting, tray and ladder systems should comply with the following standards:

740.521.1

i   Conduit systems – BS EN 61386 series
ii Cable trunking and ducting systems – the relevant part 2 of BS EN 50085
iii Tray and ladder systems – BS EN 61537.

Buried cables should be protected against mechanical damage. Conduit classified as 450N regarding protection against compression and classified as normal regarding protection against impact, according to BS EN 50086-2-4, would fulfil this requirement.

Mechanical protection should be used in all public areas and in areas where the wiring system crosses roads or walkways. Where mechanical protection is provided:

i   conduit systems should comply with BS EN 61386-21 with a classification of heavy protection against compression and heavy protection against impact. Metallic and composite conduit systems should be class 3 regarding protection against corrosion, e.g. medium protection inside and high protection outside.
ii Cable trunking and cable ducting systems should comply with BS EN 50085 series with a classification 5J regarding protection against impact.

Where subject to movement, the wiring system should be of flexible construction and should comply with BS EN 61386-23. This should use cables of type H07RNF or H07BN4-F (BS7919).

16.6.2 Cables

740.521.1 All cables should be fire rated and meet the requirements of BS EN 60332-1-2. Cables of type H07RNF or H07BN4-F (BS 7919) together with conduits complying with BS EN 61386-23 are deemed to satisfy this requirement.

Cables should have a minimum rated voltage of 450/750 V, except that, within amusement devices, cables and cords having a minimum rated voltage of 300/500 V may be used.

Where cables are buried in the ground, the route should be marked at suitable intervals and be protected against mechanical damage. Where there is a risk of mechanical damage, armoured cables or cables protected against mechanical damage due to external influence of medium severity or greater (e.g. >AG2) should be used.

16.6.3 Electrical connections

740.526 Joints should not be made in cables except where necessary as a connection into a circuit. Where joints are made, these should either use connectors in accordance with the relevant British Standards and the manufacturer's instructions or the connection should be made in an enclosure with a degree of protection of at least IP4X or IPXXD. Where strain can be transmitted to terminals the connection should incorporate cable anchorage(s).

16.6.4 External influences

740.512.2 Electrical equipment should have a degree of protection of at least IP44.

16.6.5 Switchgear and controlgear

740.51 Switchgear and controlgear should be placed in cabinets which can be opened only by the use of a key or a tool, except for those parts designed and intended to be operated
Part 2 by ordinary persons.

16.6.6 Isolation

740.537.1 It is a requirement that every electrical installation of a booth, stand or amusement device has its own means of isolation, switching and overcurrent protection; these devices should be readily accessible.

There are similar requirements for supplies to amusement devices.

Additionally, each distribution circuit should be provided with its own readily accessible and properly identified means of isolation. A device for isolation should disconnect all live conductors – line(s) and neutral conductors. Examples of devices that can be used for isolation where marked as suitable are:

i circuit-breakers
ii RCDs
iii plug and socket arrangements.

Table 53.2 **Note:** Table 53.2 of BS 7671:2008 includes guidance on device types that are suitable for isolation.

16.6.7 Luminaires

Every luminaire and decorative lighting chain should have a suitable IP rating and be securely attached to the structure or support intended to carry it. Its weight should not be carried by the supply cable, unless it has been selected and erected for this purpose. Luminaires and decorative lighting chains mounted less than 2.5 m, i.e. within arm's reach, above floor level or which could be otherwise accessible to incidental contact, should be firmly fixed, sited and guarded to prevent risk of injury to persons or ignition of materials. `740.55.1.1`

Access to the fixed light source should only be possible after removing a barrier or an enclosure, which should only be possible by the use of a tool.

Lighting chains should use H05RN-F (BS 7919) cable or equivalent, and may be used in any length provided the overcurrent protective device in the circuit is correctly rated.

Insulation-piercing lampholders must not be used unless the cables and lampholders are compatible and the lampholders are non-removable once fitted to the cable. `740.55.1.2`

Luminous tubes, signs or lamps with an operating voltage higher than 230 V/400 V a.c., e.g. neon signs, are to be installed out of arm's reach or be adequately protected from accidental or deliberate damage, to reduce the risk of injury to persons. `740.55.3`

A separate circuit should be used which should be controlled by an emergency switch. The switch should be easily visible, accessible and marked in accordance with the requirements of the local authority. Lamps in shooting galleries and other sideshows where projectiles are used should be suitably protected against accidental damage. `740.55.1.3`

Where transportable floodlights are used, they should be mounted so that the luminaire is inaccessible to non-instructed persons. Supply cables should be flexible and have adequate protection against mechanical damage. `740.55.1.4`

16.6.8 Safety isolating transformers and electronic convertors

Safety isolating transformers should comply with BS EN 61558-2-6 or provide an equivalent degree of safety. `740.55.5`

Each transformer or electronic convertor should incorporate a protective device which can be manually reset only; this device should protect the secondary circuit.

Safety isolating transformers should be mounted out of arm's reach or be mounted in a location that provides equal protection, e.g. in a panel or room that can only be accessed by a skilled or instructed person, and should have adequate ventilation.

Access by competent persons for testing or by a skilled person competent in such work for protective device maintenance should be provided. Electronic convertors should conform to BS EN 61347-2-2.

Enclosures containing rectifiers and transformers should be adequately ventilated and the vents should not be obstructed when in use.

16.6.9 Plugs and socket-outlets

740.55.7 An adequate number of socket-outlets should be installed to allow user requirements to be met safely.

In booths, stands and for fixed installations, one socket-outlet for each square metre or linear metre of wall is generally considered adequate.

Socket-outlets dedicated to lighting circuits placed out of arm's reach should be labelled according to their purpose.

When used outdoors, plugs, socket-outlets and couplers should comply with BS EN 60309-2, or, where interchangeability is not required, BS EN 60309-1.

16.7 Fire risk

GN4 Additional guidance for protection against fire for this type of location can be found in section 8.4 of Guidance Note 4: *Protection Against Fire*.

16.7.1 Luminaires and floodlights

740.55.1.5 Luminaires and floodlights should be so fixed and protected that a focusing or concentration of heat is not likely to cause ignition of any material.

16.7.2 Electric motors

422.3.7 An electric motor which is automatically or remotely controlled or which is not continuously supervised should be fitted with a manual reset protective device against excess temperature.

16.8 Electrical equipment

16.8.1 Electrical supply to devices

740.55.8 At each amusement device, there should be a connection point readily accessible and permanently marked to indicate the following essential characteristics:

i Rated voltage
ii Rated current
iii Rated frequency.

16.8.2 Electric dodgems

740.55.9 Electric dodgems should only be operated at voltages not exceeding 50 V a.c. or 120 V d.c. The circuit should have electrical separation from the electrical supply by means of a safety isolating transformer in accordance with BS EN 61558-2-4 or a motor-generator set.

16.8.3 Low voltage generating sets

740.551 It is very important that all generators are located or protected so as to prevent danger and injury to people through inadvertent contact with hot surfaces and dangerous parts.

The electrical equipment associated with the generator should be mounted securely and, if necessary, on anti-vibration mountings.

Where a generator supplies a temporary installation, forming part of a TN, TT or IT system, care should be taken to ensure that the earthing arrangements are adequate and, in cases where earth electrodes are used, they are considered to be continuously effective. In reality, this means that the drying of the ground in summer, or freezing of the ground in winter, should not adversely affect the value of earth fault loop impedance for the installation. The neutral conductor of the star-point of the generator should, except for IT systems, be connected to the exposed-conductive-parts of the generator.

542.2.2

16.9 Inspection and testing

16.9.1 The temporary installation

The electrical installation between its origin and any electrical equipment should be inspected and tested after each assembly on site.

740.6

Internal electrical wiring of roller coasters, electric dodgems, etc., is not considered as part of the verification of the electrical installation. In special cases the number of the tests may be modified according to the type of temporary electrical installation.

16.9.2 Amusement devices

The scope of Section 740 does not cover amusement devices, but, in law, there is still a requirement to ensure that the devices are fit for use.

The Amusement Device Safety Council (ADSC) is the policy-making body for safety, self-regulation and technical guidance in the UK amusement industry. This committee, in partnership with HSE, develops the policy for and oversees the Amusement Device Inspection Procedures Scheme (ADIPS). ADIPS is the fairground and amusement park industry's self-regulated safety inspection scheme which registers competent ride inspectors and the rides they inspect. The purpose of the scheme is to promote and improve fairground and amusement park safety through rules and procedures relating to the annual inspection of the amusement devices. It is supported by industry associations who require that their members use ADIPS. ADIPS is also available to any operator of amusement devices as it demonstrates 'best practice' and their compliance with the Health and Safety at Work etc. Act 1974.

Index

Guidance Note 7: Special Locations
© The Institution of Engineering and Technology

IEE Wiring Regulations and associated publications

The IEE prepares regulations for the safety of electrical installations for buildings, the *IEE Wiring Regulations* (BS 7671 *Requirements for Electrical Installations*), which have now become the standard for the UK and many other countries. It also recommends, internationally, the requirements for ships and offshore installations. The IEE provides guidance on the application of the installation regulations through publications focused on the various activities from design of the installation through to final test and then maintenance. This includes a series of eight Guidance Notes, two Codes of Practice and Model Forms for use in Wiring Installations.

Requirements for Electrical Installations BS 7671:2008 (IEE Wiring Regulations, 17th Edition)
Order book PWR1700B Paperback 2008
ISBN: 978-0-86341-844-0 **£75**

On-Site Guide (BS 7671:2008 17th Edition)
Order book PWGO170B 188pp Paperback 2008
ISBN: 978-0-86341-854-9 **£22**

Wiring Matters Magazine **FREE**
If you wish to receive a FREE copy or advertise in Wiring Matters please visit
www.theiet.org/wm

IEE Guidance Notes

A series of Guidance Notes has been issued, each of which enlarges upon and amplifies the particular requirements of a part of the IEE Wiring Regulations.

Guidance Note 1: Selection & Erection of Equipment, 5th Edition
Order book PWG1170B 216pp Paperback 2009
ISBN: 978-0-86341-855-6 **£30**

Guidance Note 2: Isolation & Switching, 5th Edition
Order book PWG2170B 74pp Paperback 2009
ISBN: 978-0-86341-856-3 **£25**

Guidance Note 3: Inspection & Testing, 5th Edition
Order book PWG3170B 128pp Paperback 2008
ISBN: 978-0-86341-857-0 **£25**

Guidance Note 4: Protection Against Fire, 5th Edition
Order book PWG4170B 104pp Paperback 2009
ISBN: 978-0-86341-858-7 **£25**

Guidance Note 5: Protection Against Electric Shock, 5th Edition
Order book PWG5170B 144pp Paperback 2009
ISBN: 978-0-86341-859-4 **£25**

Guidance Note 6: Protection Against Overcurrent, 5th Edition
Order book PWG6170B 104pp Paperback 2009
ISBN: 978-0-86341-860-0 **£25**

Guidance Note 7: Special Locations, 3rd Edition
Order book PWG7170B 144pp Paperback 2009
ISBN: 978-0-86341-861-7 **£25**

Guidance Note 8: Earthing & Bonding, 1st Edition
Order book PWRG0241 168pp Paperback 2007
ISBN: 978-0-86341-616-3 **£25**

continues overleaf ▶

Other guidance publications

**Commentary on IEE Wiring Regulations
(17th Edition, BS 7671:2008)**
Order book PWR08640
c.432pp Hardback 2009
ISBN: 978-0-86341-966-9 **£65**

Electrical Maintenance, 2nd Edition
Order book PWR05100
228pp Paperback 2006
ISBN: 978-0-86341-563-0 **£40**

**Code of Practice for In-service Inspection and
Testing of Electrical Equipment, 3rd Edition**
Order book PWR08630
152pp Paperback 2007
ISBN: 978-0-86341-833-4 **£40**

**Electrical Craft Principles, Volume 1,
5th Edition**
Order book PBNS0330
344pp Paperback 2009
ISBN: 978-0-86341-932-4 **£25**

**Electrical Craft Principles, Volume 2,
5th Edition**
Order book PBNS0340
432pp Paperback 2009
ISBN: 978-0-86341-933-1 **£25**

**Electrician's Guide to the Building
Regulations, 2nd Edition**
Order book PWGP170B
234pp Paperback 2008
ISBN: 978-0-86341-862-4 **£22**

**Electrical Installation Design Guide:
Calculations for Electricians and Designers**
Order book PWR05030
186pp Paperback 2008
ISBN: 978-0-86341-550-0 **£22**

Electrician's Guide to Emergency Lighting
Order book PWR05020
88pp Paperback 2009
ISBN: 978-0-86341-551-7 **£22**

Electrical training courses

We offer a comprehensive range of technical training at many levels, serving your training and career development requirements as and when they arise.

Courses range from Electrical Basics to Qualifying City & Guilds or EAL awards.

Train to the 17th Edition BS 7671:2008
▶ Update from 16th to 17th Edition
▶ Understand the changes
▶ New qualifying awards C&G/EAL
▶ Meet industry standards

Qualifying Courses
▶ Certificate of Competence Management of Electrical Equipment Maintenance (PAT) – 1 day
▶ Certificate of Competence for the Inspection and Testing of Electrical Equipment (PAT) – 1 day
▶ Certificate in the Requirements for Electrical Installations – 3 days
▶ Upgrade from 16th Edition achieved since 2001 – 1 day
▶ Certificate in Fundamental Inspection, Testing and Internal Verification – 3 days
▶ Certificate in Inspection, Testing and Certification of Electrical Installations – 3 days

Other 17th Edition Courses
▶ Earthing & Bonding – For designers and electrical contractors who require a good working knowledge of the E & B arrangements as required by BS 7671:2008
▶ 17th Edition Design – BS 7671 and the principles associated with the design of electrical installations

To view all our current courses and book online, visit
www.theiet.org/coursesbr

To discuss your training requirements and for on-site group training, please speak to one of our advisors on +44 (0)1438 767289

For more information, visit www.theiet.org/wiringregs

Order Form

How to order

BY PHONE:
+44 (0)1438 767328
BY FAX:
+44 (0)1438 767375
BY EMAIL:
sales@theiet.org
BY POST:
The Institution of
Engineering
and Technology,
PO Box 96,
Stevenage
SG1 2SD, UK
OVER THE WEB:
www.theiet.org/books

*Postage/Handling: Postage within the UK is £3.50 for any number of titles. Outside UK (Europe) add £5.00 for first title and £2.00 for each additional book. Rest of World add £7.50 for the first book and £2.00 for each additional book. Books will be sent via air-mail. Courier rates are available on request, please call +44 (0) 1438 767328 or email sales@theiet.org for rates.

** To qualify for discounts, member orders must be placed directly with the IET.

GUARANTEED RIGHT OF RETURN:
If at all unsatisfied, you may return book(s) in new condition within 30 days for a full refund. Please include a copy of the invoice.

DATA PROTECTION:
The information that you provide to the IET will be used to ensure we provide you with products and services that best meet your needs. This may include the promotion of specific IET products and services by post and/or electronic means. By providing us with your email address and/or mobile telephone number you agree that we may contact you by electronic means. You can change this preference at any time by visiting www.theiet.org/my.

Details

Name:

Job Title:

Company/Institution:

Address:

Postcode: Country:

Tel: Fax:

Email:

Membership No (if Institution member):

Payment methods

☐ By **cheque** made payable to The Institution of Engineering and Technology

☐ By **credit/debit card:**

☐ Visa   ☐ Mastercard   ☐ American Express   ☐ Maestro Issue No:_____

Valid from: ☐☐ ☐☐   Expiry Date: ☐☐ ☐☐   Card Security Code: ☐☐☐☐
(3 or 4 digits on reverse of card)

Card No: ☐☐☐☐ ☐☐☐☐ ☐☐☐☐ ☐☐☐☐

Signature_____ Date _____
(Orders not valid unless signed)

Cardholder Name:

Cardholder Address:

Town: Postcode:

Country:

☐ By official **company purchase order** (please attach copy)
EU VAT number:_____

Ordering information

Quantity	Book No.	Title/Author	Price (£)
		Subtotal	
		- Member discount**	
		+ Postage /Handling*	
		+ VAT (if applicable)	
		Total	

Membership

Passionate about engineering? Committed to your career?

Do you want to join an organisation that is inspiring, insightful and innovative?

One of the most highly recognised knowledge sharing networks in the world, membership to the Institution of Engineering and Technology (IET) is for engineers and technologists working or studying in an increasingly multidisciplined, digital and global environment.

Joining the IET and having access to tailored products and services will become invaluable for your career and can be your first step towards professional qualifications.

You could take advantage of ...

▶ 18 issues per year of the industry's leading publication, *E&T* magazine.

▶ Professional development and career support services to help gain professional registration.

▶ Discounted rates on dedicated training courses, seminars and events covering a wide range of subjects and skills.

▶ Watch live IET.tv event footage at your desktop via the internet, ask the speaker questions during live streaming and feel part of the audience without physically being there.

▶ Access to over 100 local networks around the world.

▶ Meet like-minded professionals through our array of specialist online communities.

▶ Instant online access to over 70,000 books, 3,000 periodicals and full-text collections of electronic articles – wherever you are in the world.

▶ Discounted rates on IET books and technical proceedings.

Join online today at www.theiet.org/join or contact our membership and customer service centre on +44 (0)1438 765678

Professional Registration

What type of registration is for you?

Engineering Technicians (EngTech) are involved in applying proven techniques and procedures to the solution of practical engineering problems. You will carry supervisory or technical responsibility, and are competent to exercise creative aptitudes and skills within defined fields of technology. Engineering Technicians also contribute to the design, development, manufacture, commissioning, operation or maintenance of products, equipment, processes or services.

Incorporated Engineers (IEng) maintain and manage applications of current and developing technology, and may undertake engineering design, development, manufacture, construction and operation. Incorporated Engineers are engaged in technical and commercial management and possess effective interpersonal skills.

Chartered Engineers (CEng) develop appropriate solutions to engineering problems, using new or existing technologies, through innovation, creativity and change. They might develop and apply new technologies, promote advanced designs and design methods, introduce new and more efficient production techniques, marketing and construction concepts, pioneer new engineering services and management methods. Chartered Engineers are engaged in technical and commercial leadership and possess interpersonal skills.

For further information on Professional Registration (EngTech/IEng/CEng), tel: +44 (0)1438 765673 or email: membership@theiet.org